THE GENESIS QUEST

The Genesis Quest

The Geniuses and Eccentrics
on a Journey to Uncover
the Origin of Life on Earth

MICHAEL MARSHALL

The University *of* Chicago Press

The University of Chicago Press, Chicago 60637

Published 2020
Paperback edition 2021
Printed in the United States of America

30 29 28 27 26 25 24 23 22 21 1 2 3 4 5

ISBN-13: 978-0-226-71523-0 (cloth)
ISBN-13: 978-0-226-81804-7 (paper)
ISBN-13: 978-0-226-71537-7 (e-book)
DOI: https://doi.org/10.7208/chicago/9780226715377.001.0001

First published in Great Britain by the
Orion Publishing Group, London, 2020.

Library of Congress Cataloging-in-Publication Data

Names: Marshall, Michael (Science writer), author.
Title: The genesis quest : the geniuses and eccentrics on a journey to uncover the origin of life on earth / Michael Marshall.
Description: Chicago : The University of Chicago Press, 2020. | Includes bibliographical references and index.
Identifiers: LCCN 2020013379 | ISBN 9780226715230 (cloth) | ISBN 9780226715377 (ebook)
Subjects: LCSH: Life—Origin—Research—History. | Scientists.
Classification: LCC QH325 .M2995 2020 | DDC 576.8/3—DC23
LC record available at https://lccn.loc.gov/2020013379

♾ This paper meets the requirements of ANSI/NISO Z39.48-1992 (Permanence of Paper).

For Sarah and Libbet

CONTENTS

PART THREE
Scattered, Divided, Leaderless

PART FOUR
Reunification

INTRODUCTION

How did life begin? Why are we here? These are among the most profound questions we can ask. They are about both ourselves and our relationship to the wider universe.

Yet while the question of how life started on our planet is an obvious one, scientists only really began engaging with it in the early twentieth century. Research into the origin of life is barely a century old. Even today, there are only a few dozen laboratories that directly tackle this important subject.

This book is the story of the scientists who have attempted to explain how and why life arose on our planet. It explores all the major ideas, explains how the scientists came to develop them, and delves into their strengths and weaknesses. It is crucial to tackle all the ideas because, despite what their proponents say, it is clear that most of them cannot be correct. It is only by trying them all on for size that we can start to see how life might really have begun. In recent years, origins researchers have started to devise a kind of 'grand unified theory' that has a good chance of being correct – in part because it includes the best elements from the older ideas. It is only by examining what has gone before that we can unlock the secrets of what might come next.

Our story begins in the 1920s, when after decades of minimal progress a scenario was proposed that gained widespread acceptance: that life began in the famous 'primordial soup'. However, little progress was made for decades, until a seminal experiment performed by Stanley Miller in the early 1950s seemed to show that the chemicals of life could have formed naturally on the young Earth.

The Miller experiment kick-started a new field of science called prebiotic chemistry, which sought to make life's building blocks from simple chemicals. However, the problem of life's origin swiftly became immeasurably more complicated, thanks to the emerging knowledge of the complex workings of living cells. So intricate and interdependent is the machinery inside even the simplest bacterium, it is hard to imagine what a basic, primordial version would look like. Strip out even one of the key systems, and the organism dies.

As a result, from the 1960s on, several competing ideas arose, each championed by a faction of scientists for the remainder of the twentieth century. Each of these new ideas focused on one key component of living cells, which was supposed to arise on its own and then bring the others into being as necessary. For example, one popular idea known as the 'RNA World' holds that a simple genetic molecule came into being and found a way to copy itself. Unfortunately, all these ideas ultimately fall down. None of life's components, on their own, can achieve living status.

However, in the twenty-first century a few researchers have begun exploring ways to build living cells with all the key systems intact, but in hugely simplified form and using a bare handful of chemicals. This new approach, which was long viewed as virtually inconceivable, has had an astonishing run of success. The earlier ideas are still much discussed, but the new hypothesis has a better chance of being correct. It now seems that life began, not with a single component like a gene, but with several components that could work together. Life is less about a particular substance, and more about the way a group of substances behave when they are combined.

The quest to understand how life began is also the attempt to answer a profound question: is life inevitable? That is, was life always going to form given how the universe works, or is it a freak accident?

In his 1970 book *Chance and Necessity*, the French biochemist Jacques Monod staunchly argued that the universe is not shaped for life like us.[1] He contended that the origin of life on Earth was stupendously unlikely. Life may only have formed once, here, in the entire history of the universe. 'The universe was not pregnant with life nor the biosphere with man,' he asserted. 'Is it surprising that, like the person who has just made a million at the casino, we should feel strange and a little unreal?' Monod's attitudes were evidently shaped by existentialist philosophers like Jean-Paul Sartre and Albert Camus, who wrote of their nausea and alienation at a life lived without the moral compass offered by a god. He argued that we should all now feel 'alone in the unfeeling immensity of the universe'.

Monod's high-flown rhetoric has impressed a lot of people over the years, but aside from his portentous prose he offered precious little evidence. His main argument was that the evolution of new species is ultimately driven by random changes in genes. This meant that the sole driving force of evolution was 'pure chance, absolutely free but blind'. However, Monod arguably went too far. It is true that genetic mutations are random, but which ones survive and become common in a population is not necessarily random at all. Instead, the mutations that survive are those that bestow benefits on the organisms that carry them, or at least allow the genes to become more common. That is why certain traits have evolved more than once, with different genetic underpinnings: for example, flight arose independently in insects, birds and bats. Monod was looking at the history of life through random-tinted glasses. It is a mistake to overemphasise the randomness of evolution and neglect its many regularities.

Let's now consider the opposite viewpoint. Origin-of-life researchers often say that life was somehow predestined. This can come across as semi-mystical, but there is at least some basis for thinking it is true.

The person who argued most fiercely that life was inevitable was the Belgian cell biologist Christian de Duve.[2] In his book *Vital Dust*, he argued that life is 'a cosmic imperative'.[3] Given how complex living organisms are, de Duve argued, they are stupendously unlikely to have arisen by chance. 'We are being dealt thirteen spades not once but thousands of times in succession,' he wrote. 'This is utterly impossible, unless the deck is doctored.' In other words, life must be easy to form under the right circumstances, or we would not be here. Over the course of the story, we will see that there are many natural processes that favour the formation of the chemicals of life, and of lifelike structures – which supports de Duve's contention.

Ultimately, de Duve's arguments rely on a rule of thumb called the Copernican Principle, or the Mediocrity Principle. The basic message is: assume we are ordinary. Scientists have tried to stick to this ever since astronomer Nicolaus Copernicus presented evidence that the Earth orbits the Sun, and is not at the centre of the universe as almost everyone had assumed. The Copernican Principle suggests that the chemicals and environments found on Earth are probably typical of rocky planets. In that case, life is fairly likely to pop up on any reasonably Earth-like planet in the right sort of orbit around the right sort of star.

However, we should not take this too far. Look again at the last sentence of the previous paragraph, and count the caveats. Life may be likely, but de Duve knew that does not mean it is common – and a glance at the known universe tells us that hardly any of it is remotely suitable for life.

Consider the Earth, the only place known to support life. Most of it is utterly inhospitable. The habitable zone is at most a few tens of kilometres thick, spanning the outer layers of rock, the seas and the lower layers of the atmosphere. Go too high up and the air becomes too thin, the radiation from the Sun too intense, and the temperatures too extreme. Meanwhile, go too far underground and the temperature and pressure become lethal. Our world may be a living one, but over 90 per cent of its mass is dead.

Beyond Earth, things get even sketchier. Even if trillions of planets are home to life, there are still vast volumes of empty space in every solar system, not to mention all the interstellar space and the unimaginable voids between the galaxies. Some of these hollow spaces are hundreds of millions of light-years across. Unless there is life within stars or in the empty blackness of intergalactic space, the universe is mostly dead. It is a good bet that 99 per cent of the volume of the universe is lifeless, as is 99 per cent of matter.

It is stretching the definition of 'favoured' quite far to claim that this is a universe that especially favours life. In fact, the whole argument can be turned around: our universe is egregiously hostile to life in all but a few minuscule patches. If the universe is fine-tuned for anything, it's for empty space, dust, stars and planets. But on at least some of those planets, life is quite likely. Life is an edge phenomenon in the cosmos, something that has snuck in.

This perspective is somewhere between the two extreme positions staked out by Monod and de Duve – albeit closer to de Duve. The universe does not look like an engine optimised for creating life, but nor is it wholly bent on exterminating it. Instead, it is largely occupied with other things. Over the course of the story, we will discover special places on the Earth that seem to be ideal incubators for life – and see that those places are rare in the wider universe.

As well as a scientific odyssey, this is a story with a fascinating and often brilliant cast of characters, from the rambunctiously eccentric J. B. S. Haldane to the irascible Günter Wächtershäuser. Many of the people involved have been Nobel Prize winners, and they include some of the most brilliant minds of the last century. Some, like Carl Sagan, are well known; others will be less familiar. Discovering this unlikely group of pioneers is one of the great pleasures of the story.

Sadly, astute readers will soon notice that most of the scientists involved are male. The first three chapters do not contain contributions by a single female scientist. This streak does not break until Rosalind Franklin makes her appearance in chapter 4, helping to reveal the structure of DNA. Even after that, almost all the key figures are male. This should not come as too much of a surprise, as the institutions of science have been as beset by sexism as the rest of society – as science journalist Angela Saini has exposed in her book *Inferior*.[4] I have tried to find female origins researchers who have been overlooked, with little success, and have sought to name female co-authors alongside the male heads of lab. But it would be misleading to make the story look more egalitarian than it was. For what it's worth, I hope that future accounts of origins research, written when more time has elapsed, will feature a more diverse cast of characters.

Despite these issues of representation, the story of the quest to understand genesis is a universal one, in which everyone can find pleasure and fascination. By asking how life came to be, we are implicitly asking why we are here, whether life exists on other planets, and what it means to be alive. This book is the story of a group of fragile, flawed humans who chose to wrestle with these questions. By exploring the origin of life, these people have caught a glimpse of the infinite.

Michael Marshall,
Devon, February 2020

PART ONE

Primordial Science

'Most chemists believe, as do I, that life emerged spontaneously from mixtures of molecules in the prebiotic Earth. How? I have no idea.'

George Whitesides, in his 2007 address accepting the American Chemical Society's Priestley Medal for distinguished service[1]

Chapter 1

The Biggest Question

Everywhere you look, you are looking at life. This is true even if you stare at a brick wall, at a computer screen or straight up into the sky. Even if there are no big animals or plants in sight, there are insects, microscopic animals and, always, uncounted billions of bacteria. There is life on every surface of our little planet, even in seething hot geothermal ponds, in the crushing pitch-black depths of the sea and above the clouds where the air grows thin.

But it was not always so. There was a time when Earth was dead. A time when it was a ball of semi-molten rock, blasted by pummelling meteors and dotted with violently erupting volcanoes. There were no seas then, and no oxygen to breathe in the air. If we were to somehow travel back in time to visit this young Earth, without protective clothing and an oxygen tank, the only question would be how we would die first: asphyxiation or simply being incinerated.

Somehow, our planet transformed itself from a lifeless hellscape to the green-and-blue paradise our species is currently in the process of screwing up. What happened back then? What was the first living thing: what did it look like, what was it made of, and how did it come to be?

This question taps into another mystery, one that has boggled our minds for centuries: are we alone in the universe? So far, Earth is the only world known that supports life. What conclusion should we draw from that? Perhaps the formation of life is overwhelmingly likely, given the right conditions, in which case the cosmos must surely be teeming with life – and the other worlds in the Solar System just didn't fit the bill. Or maybe life is staggeringly unlikely, something seemingly miraculous that happens only on one world in a billion billion billion. In that case we are alone, existentially so, and when we look up at the stars we are looking at emptiness.

Humans have told stories about the creation of life for thousands of years. These myths can be beautiful and meaningful, but none of them really explains how life could have formed solely from non-living building blocks. To illustrate this, consider the Norse creation myth, which was splendidly retold by Neil Gaiman in *Norse Mythology*.[1] Gaiman tells of enormous glaciers that extend into a land of volcanoes and fire. Where the ice is melted by the fire, a huge person appears called Ymir, who is the ancestor of all giants. Alongside Ymir is a vast hornless cow, who feeds by licking the salty ice. Ymir in turn guzzles the cow's milk and grows.

This is a vivid story, but it doesn't take much pedantry to see that it doesn't explain anything. If you melt some ice you don't normally get a giant, let alone a monumental cow. The story skips the difficult part in order to get to the resonant bit, which comes a page later when the god Odin and his brothers murder Ymir, and in so doing create the world.

At least the Norse myth explicitly depicts life coming from non-life. Other creation myths cheat even more spectacularly by saying that life was created by pre-existing life, the origin of which is left as an exercise for the reader. Any story in which life is created by a god or gods is guilty of this. The retort 'but where does the god come from?' is so obvious it can seem childish to utter it, but being obvious doesn't make it an invalid question.

Once you realise that creation myths don't offer an answer, it becomes clear that for most of human history nobody was really asking how life began. This seems to be because people had two implicit ideas about the nature of life, which between them closed off the question.

The first assumption was that there was something special or magical about living matter that made it fundamentally different to non-living matter. This idea is called 'vitalism' and it recurs across cultures. It's right there in the Book of Genesis: 'And the Lord God formed man of the dust of the ground, and *breathed into his nostrils the breath of life* [my italics]; and man became a living being.' The Stoic philosophers of ancient Greece wrote about *pneuma*, which means 'breath' but also implies 'spirit'. Similarly, Aristotle discussed *psyche*, which means something like 'soul' but is not necessarily conscious or intelligent. If you believe you have a soul, maybe made of some nebulous 'energy' that will leave your body when you die, then you believe in vitalism.

There is something profoundly attractive about vitalism. It seems obvious to everyone that an elephant is not the same thing as a rock, and that difference is not just fine detail but something fundamental. Life is special.

But while vitalism is intuitive, it is also plain wrong. There is no life force – or if there is, nobody has been able to detect it or even specify what it is. Instead, much of the last 200 years of biology have been about explaining the unique properties of living things in terms of the non-living chemicals from which they are made.

The story that is often told about vitalism is that it was disproved by the German chemist Friedrich Wöhler. Historians of science have argued that this is something of a myth, as we'll see, but it remains a crucial episode.

By the early 1800s, scientists had identified a number of chemicals that seemed to be unique to life. They were only found in living things, and nowhere else. One such chemical was urea, which is found in urine and gives it its yellow-brown colour.

Along came Wöhler. In 1824 he was puzzled by some white crystals that formed during a chemical experiment. Four years later, he identified them as urea.[2] Crucially, he also reported that he had made urea from ammonium chloride: a chemical that has nothing to do with life.

However, it is not clear that this really disproved vitalism, or that Wöhler saw it in that light. Certainly, Wöhler's result was bad for vitalism. If it was possible for one of the chemicals of life to be made from a chemical that has nothing to do with life, presumably others could be. However, Wöhler's own discussion didn't really put the boot into vitalism at all. Instead, the idea that Wöhler disproved vitalism gradually emerged in commentaries written by other scientists over the following few decades. This process has been traced by Peter Ramberg, a historian of science, who calls it the 'Wöhler myth'.[3] By the 1930s, one can find descriptions of Wöhler trying over and over to synthesise urea from other chemicals, determined to refute vitalism. This does not seem to be how Wöhler saw it.

Setting aside the historical controversy, there is a better reason to reject vitalism. Put simply, vitalism is an atrocious explanation for the existence of life. It does nothing to explain what is special about living things, but merely puts a label on that specialness. If you posit the existence of an undetectable and undefined energy that can turn non-living matter into living organisms, it doesn't explain the existence of life. It just raises the question of what this energy is, how it is generated, and how it can affect the matter in living organisms without being detected by any sensor or experiment.*

Let's consider also what it would mean if there really were a life energy or vital substance. In theory, we ought to be able to extract this *élan vital* from a living organism – killing the organism in the process – and inject it into something non-living like a rock or a teddy bear. This inanimate object would then come to life. It hardly needs saying that living teddy bears don't exist, outside of stories like *Winnie-the-Pooh* and *Akira*, which suggests things don't work like that.

Finally, the idea of a life force falls foul of Occam's razor: the rule of thumb that one should explain mysteries using as few assumptions as possible. We should only invoke a new form of energy if we really have to, if all attempts to explain life using known phenomena have failed. The enormous progress made

* This is also the reason not to accept 'god(s) did it' as the explanation for the existence of life: it explains nothing. If you asked how a car was made and were told 'someone built it', you would not be satisfied.

over the last few centuries in our understanding of life's inner workings suggests we do not need to assume the existence of anything truly new.

Nevertheless, vitalism has proved peculiarly difficult to shake. It appeals to our deepest intuitions, even when we should know better. As late as 1913, the English biochemist Benjamin Moore was advocating for 'biotic energy', which was little more than a rebranding of vitalism. In his book *The Origin and Nature of Life*, Moore drew an analogy with the then newly discovered phenomenon of radioactivity, arguing that if atoms could possess 'new, strange forms of energy', so too could living matter.[4] The only answer to this special pleading is to ask for direct evidence of these 'new energy properties'. None has ever been forthcoming.

However, it seems clear why vitalism holds such instinctive appeal. We all have a sense that there is something special and precious about life, and we are reluctant to accept any idea that threatens to take that away. There is something seemingly cold and dehumanising about the idea that there is nothing remarkable about living matter. The thing is, life energy is not the only way we can imagine living matter as special. Modern science strongly suggests that living matter is made up of exactly the same atoms as non-living objects like rocks and scented candles. Instead, what is special is the way these atoms are arranged, and in particular the patterns of motion they perform. An elephant may contain nothing more than carbon and a few dozen other elements, but nobody could predict the peculiar wisdom of elephant matriarchs simply by looking at that list of chemicals. We have to look at life through a different set of lenses to appreciate what it is. The computer scientist Steve Grand put it best in his book *Creation*, which documents his attempt to make artificial life in a computer: 'Life is more than just clockwork, even though it is nothing but clockwork.'[5]

Today the idea of a life force is common in alternative medicine. It's often dressed up as ancient wisdom, such as the Chinese concept of *chi*, which is supposed to underlie acupuncture. Oddly enough, vitalism also clings on in science fiction. The Time Lords in *Doctor Who* possess 'regeneration energy', which allows them to reshape their bodies after suffering a terminal injury. The show presents this idea literally: in the 2013 episode 'The Time of the Doctor', the Doctor has run out of regeneration energy and is close to death, until he receives a top-up that allows him to regenerate. *Doctor Who* is not exactly known for its scientific accuracy, but the much harder-edged *Babylon 5* also used the trope, in the form of a machine that could transfer life energy from one person to another. These ideas have a veneer of futurism, but in fact they are deeply primitive.

Even if some people saw through vitalism, there was a second reason for science to ignore the origin of life. It was widely believed that life forms all the time, from dead matter. This idea of 'spontaneous generation' is another one that recurs across many cultures, from Christianity ('And the earth brought forth grass, the herb that yields seed according to its kind, and the tree that yields fruit') to Chinese scriptures. Often a god of some kind was involved, but not always.

It's easy to see why people would think that life can spontaneously form from non-living matter. Leave a piece of meat in a warm place for a few days and maggots will form in it, and unless you keep a close watch you will not see their insect parents laying their eggs in it. It seems as though the maggots have formed of their own accord, given only a food source to lure them into existence. Almost all things go mouldy or rot if you leave them long enough, so it was natural to think that life was constantly being formed. Aristotle spelled it out in *History of Animals*: 'of these instances of spontaneous generation some come from putrefying earth or vegetable matter, as is the case with a number of insects, while others are spontaneously generated in the inside of animals out of the secretions of their several organs'.

Vitalism and spontaneous generation are arguably mutually contradictory. If living organisms form in mud, from where do they get their life force? It can't come from the mud, because the mud isn't alive and therefore doesn't have any. Still, historically people have found ways to believe both at once.

However, by the middle of the nineteenth century spontaneous generation was under attack from scientists. By this time biologists had discovered the life cycles of parasitic worms, which were previously thought to appear in human intestines by spontaneous generation. At a stroke this removed one of the key arguments for spontaneous generation, because instead of the worms appearing from nowhere, they in fact had parents. There were also rumbling controversies about the nature of diseases like cholera – which we now know to be caused by microorganisms – and about the processes of decay and fermentation.

The idea had also become politically charged. For centuries the Christian Church backed spontaneous generation: both Saint Augustine and Thomas Aquinas wrote lengthy tracts fitting it into their theology. Augustine thought God had endowed the universe with the potential for life, which continually sprang forth, while Aquinas saw each new emergence of life as another divine miracle. However, by the seventeenth century the dogma had shifted and the Church rejected spontaneous generation. As a result, belief in spontaneous generation became associated with atheism, and ultimately became tied up with the anticlerical and liberal ideas of the French Revolution.[6]

Into this breach stepped the French naturalist Félix-Archimède Pouchet, the director of the Natural History Museum at Rouen. In 1858, Pouchet published a paper in which he described experimental evidence for spontaneous generation.[7] Pouchet placed hay in water and left it to infuse, making a kind of hay tea. He then boiled it to kill off any microorganisms living in it. Finally, he exposed the liquid to purified air, which should also be sterile. All of this was done under a layer of liquid mercury, so that no microorganism could drift down into the liquid.

Despite these precautions, the surface of the tea went mouldy. Pouchet claimed that this mould formed by spontaneous generation, since he had removed all potential sources of life. The following year he published a book arguing his case: *Heterogenesis, or Treatise on Spontaneous Generation*.[8]

This did not go over well with the French Academy of Sciences. They established a prize of 2500 francs, to be awarded to 'him who by well-conducted experiments throws new light on the question of so-called spontaneous generation'. The 'so-called' is telling: the academy was unsympathetic to Pouchet's claims.

The competition drew the attention of the biologist Louis Pasteur. He was still in the early stages of his career: it would be many years before he discovered the principles of vaccination. Instead, Pasteur had spent the latter half of the 1850s studying the fermentation of lactic acid: the chemical process that turns milk sour. Whereas several chemists had argued that this was a purely chemical process, Pasteur showed that microorganisms were crucial. This ultimately led to the realisation that foods like milk can be preserved by heating them to eliminate microorganisms. We call this pasteurisation.

Pasteur was convinced that the microorganisms responsible for fermentation float around in the air. Pouchet's experiment seemingly disproved this, but Pasteur argued that Pouchet hadn't been careful enough. He suggested, forcefully, that there were microorganisms in the dust sitting atop the mercury layer, and that these had contaminated the hay tea. So he tried a different approach.

Pasteur mixed sugared water, yeast, urine and beet juice in a flask, then heated the neck of the flask and extended it into the shape of an S, like the neck of a swan. This swan-neck arrangement would allow air in, but trap any microorganisms that entered. Pasteur then boiled the liquid and watched as absolutely no mould formed. Then, in a dramatic *coup de grâce*, Pasteur snapped the neck off the flask and dipped it in the boiled liquid. Only then did microorganisms appear, having finally made it into the liquid.

In a second experiment, Pasteur sealed the necks of his flasks and carried them up into the French Alps, where the thin air presumably contained fewer microorganisms. Once there, he opened the necks to let the air in. The higher he went, the less likely the liquids were to go mouldy.

The academy duly awarded Pasteur his prize in 1861. Spontaneous generation had been disproved.

Science textbooks often present the Pasteur–Pouchet debate as a dramatic turnaround – so much so that revisionist historians of science sometimes argue that Pasteur got lucky or was unduly biased. In truth, spontaneous generation had been losing ground for decades, and the argument rumbled on for some time. The British scientist Henry Bastian presented experiments that seemed to show spontaneous generation throughout the latter half of the 1800s, but they were discredited by the discovery that some bacteria could form heat-resistant spores that survived Bastian's methods of sterilisation.

At any rate, once Pasteur had won the debate, the origin of life should have become a live scientific question. But it didn't. Few researchers tackled it.

This was despite the problem having gained a new urgency thanks to a major discovery. In 1859, the same year that Pouchet published his book and unwittingly unleashed Pasteur, the British naturalist Charles Darwin had published his magnum opus, *On the Origin of Species*.[9] In it, Darwin, inspired by his experiences sailing round the world, set out his theory of evolution by natural selection, which explains the amazing diversity of life on Earth. Other researchers had already suggested that each species was not created individually by God and set in stone forever more, as was thought. Darwin bolstered the case by offering an explanation. His theory of natural selection stated that species gradually changed in response to the environments in which they lived, occasionally giving rise to entirely new species. Every kind of living thing, from the largest blue whale to the tiniest bacterium, was ultimately descended from a single common ancestor that lived long ago. As Darwin would later spell out, that included us: humans were animals, descended from ape-like ancestors.

Darwin's theory of evolution caused the most almighty hullabaloo. Many Christians were apoplectic: Darwin seemed to be cutting God out of the creation of life altogether and proposing a wholly atheistic explanation for the wonders of nature. In fact, Darwin was rather hesitant on this point and his writings barely touch on religion. But nevertheless his ideas were a body blow for the supremacy of religious belief of any kind. The incredible intricacy of animals and plants looks like it has been designed, which implies a designer, but Darwin showed that there was no need to invoke any such creator. Even startlingly complex organs like the human eye arose gradually, bit by bit, from simple beginnings.

Nowadays science takes evolution as an accepted fact, but the row over whether it was true rumbled on for the rest of the nineteenth century and well into the twentieth. Not until the 1930s and 1940s, when evolutionary theory was integrated with the emerging science of genetics and heredity in what became known as the 'modern synthesis', was the question firmly settled. For many religious people today, of course, it remains a disputed idea or even an outright lie. The point is simply that biologists had quite enough to do, what with defending evolution against the religious while figuring out exactly how it worked, to worry about how the whole thing might have started.

Contrary to popular belief, evolution says nothing about how life began. It assumes that all living things are descended from a common ancestor, but where that ancestor came from is beyond the scope of the theory. In a sense, Darwin's theory created the question of life's beginning, which did not previously exist in people's minds. As Bill Mesler and H. James Cleaves II put it in *A Brief History of Creation*: 'People asked where the first monkey came from, or the first shark . . . but not the first *species*, period.'[10] Those who believed in spontaneous generation thought it gave rise to many different kinds of life, from flies to worms. Darwin's theory instead suggested there was a single living organism that came first, and which was the ancestor of everything else.

Darwin did not discuss the origin of life in his books, but he did mention it in a letter to the botanist Joseph Hooker written in 1871.[11] This unassuming little note is totemic, for in it Darwin speculated that life could have formed naturally on Earth in the distant past, given the right conditions. Darwin wrote:

> It is often said that all the conditions for the first production of a living organism are now present, which could ever have been present.— But if (& oh what a big if) we could conceive in some warm little pond with all sorts of ammonia & phosphoric salts,—light, heat, electricity &c present, that a protein compound was chemically formed, ready to undergo still more complex changes, at the present day such matter w[d] be instantly devoured, or absorbed, which would not have been the case before living creatures were formed.

In other words, suppose that millions of years ago, when Earth was lifeless, a pond formed. It contained a mix of simple chemicals and was bathed in sunlight. The chemicals in the water might eventually form a protein, one of the crucial molecules of life, and that protein might then become ever more complex until it could be said to be alive. This could never happen today, because any protein that formed in the water would be instantly eaten by some hungry organism. But with no life around, life's chemicals could develop in peace.

Darwin's idea is a sketchy one, as you would expect of

something that was dashed off in a short letter that – as he went on to note – was written while one of his daughters was suffering a severe case of measles. But it is nevertheless arguably the first modern suggestion for how life could have begun. It implicitly asks the key question: how do lifeless chemicals assemble themselves in such a way that they become a living thing?

Let's narrow down the problem. What is the simplest form of life we can imagine? Where might it have formed – and when?

On the face of it, imagining what the first living organism might have looked like seems a tall order. Life is so ridiculously, amazingly diverse. Consider a blue whale, the largest animal ever to exist at nearly thirty metres long, swimming majestically through the ocean. It has a huge brain, a heart the size of a car, a mouth filled with baleen for sieving tiny krill out of the water, and a set of imposing genitals. And now consider fly agaric, a red mushroom with white spots that looks like it should have a tiny door in its stalk with a fairy poking its head out. A few centimetres in size, it has none of the organs a blue whale does. It does not move and instead simply pokes up from the floor of a forest. The two organisms appear to have almost nothing in common, but in fact they are both built of the same microscopic structures.

The first clue to life's underlying unity was discovered by the British scientist Robert Hooke in the seventeenth century. Hooke was a grumpy genius, one of the founders of the scientific revolution, a rival to Isaac Newton, and one of the first members of the Royal Society. Like many early scientists, he ranged across many fields, from mechanics – Hooke's law of elasticity is named for him – to timekeeping and astronomy. But perhaps his greatest contribution was his 1665 book *Micrographia*, the Royal Society's first major publication.[12]

In *Micrographia*, Hooke described what he had seen when he looked at a host of objects under a compound microscope. Although microscopes had existed for several hundred years, they had been drastically improved in recent decades, and Hooke was seeing things that no human had ever seen before. He illustrated the book lavishly, which may account for its immense impact.

In one section, Hooke described what he saw when he looked at thin slices of bottle cork under his microscope. The cork was peppered with innumerable tiny chambers or pores. They reminded Hooke of the tiny rooms in which monks lived, and so he named them 'cells' after those chambers.

Meanwhile, in the city of Delft in the Dutch Republic, Antonie van Leeuwenhoek was also experimenting with microscopes. He had found a way to make particularly high-quality lenses, and as a result saw even more detail than Hooke. His most crucial find was communicated in a letter he sent to the Royal Society in October 1676. It was published the following year after a great deal of internal debate and controversy – in which Hooke firmly defended Leeuwenhoek after confirming some of his observations for himself.[13]

Leeuwenhoek's opening sentences set forth an extraordinary claim. In 1675 he had discovered 'living creatures in Rain water' that had been standing for a few days in an earthen pot. The tiny creatures were ten thousand times smaller than the smallest critters yet seen. Over eleven pages, Leeuwenhoek enumerated the different shapes and behaviours of these tiny 'animalcules' or 'living Atoms'. Some were oval, but occasionally became near-perfect spheres. Others were 'twice as long as broad' and seemed to have 'little feet, whereby they moved very briskly'.

Leeuwenhoek had discovered microorganisms: living things far smaller than even the tiniest of insects and mites. What's more, he claimed they were everywhere: in pond water, in the sea, on the surfaces of objects, even floating in the air. It is no wonder other scientists were sceptical. But Leeuwenhoek's observations were unimpeachable, and soon the science of microbiology would begin in earnest. Later researchers would group the animalcules into different classes, giving them names like '*Amoeba*' and 'bacteria' that are still in use today. But the crucial realisation was that they were all made of cells, much like those Hooke had seen. These cells came in different shapes and sizes, and some had armour plating or things that looked like spinning, whipping tails. But, at root, each animalcule was a single cell.

Over the next 150 years, cells began turning up everywhere. As biologists examined living tissue under the microscope, they kept finding cells. Brain cells, blood cells, liver cells, muscle cells, single nerve cells that stretched for over a metre without a break, swollen egg cells and frantically swimming sperm cells: there seemed to be almost nothing in the bodies of humans and other organisms that was not made of cells.

In 1838, the German botanist Matthias Schleiden put down in writing what many biologists quietly suspected but had not quite spelled out. In a paper with the riotous title 'Contributions to Our Knowledge of Phytogenesis', Schleiden described the anatomy and growth of plants and argued that cells are the foundation of everything that makes a plant a plant.[14] As he put it, 'the entire growth of the plant consists only of a formation of cells within cells'.

At the same time, another German, the biologist Theodor Schwann, was examining animal cells under the microscope. Supposedly, the two men dined together and discussed their work – prompting Schwann to realise the similarities between the plant cells Schleiden studied and the animal cells he had been looking at. The result was 1839's *Microscopical Researches into the Accordance in the Structure and Growth of Animals and Plants*, in which Schwann took the obvious next step and declared that all living things, not just plants, were fundamentally made up of cells.[15] This, then, is what links blue whales and fly agaric mushrooms despite all their differences: they are both made of cells.

'How broad is the distinction between a muscle and a nerve, between the latter and cellular tissue (which agrees only in name with that of plants), or elastic or horny tissue, and so on,' Schwann wrote, bending over backwards to show that he knew what a big claim he was making. 'When, however, we turn to the history of the development of these tissues, it appears that all their manifold forms originate likewise only from cells, indeed from cells which are entirely analogous to those of vegetables, and which exhibit the most remarkable accordance with them in some of the vital phenomena which they manifest.'

The idea was that all living matter is either made of cells or, in the case of things like nails and feathers, produced by them. This became known as cell theory and today it is beyond question.

How might one cell give rise to another? This question was gradually solved over the course of the 1830s after a string of researchers saw the process in their microscopes.[16] It was startlingly simple: cells could divide themselves into two. A spherical cell would slowly pinch itself in the middle, the central part becoming narrower until the cell looked like a bow tie. Finally the last remaining link vanished, and instead of a single 'mother' cell there were two 'daughter' cells.

The final touch was put in place by the Prussian-born biologist Robert Remak.[17] He realised that this process of cell division was the only way that new cells are formed. During the 1840s and 1850s he published a stream of evidence to this effect, but was widely disbelieved and – as a result of his Jewish faith – repeatedly denied the professorship he had so clearly earned. His ideas were finally popularised by Rudolf Virchow, who summed them up in 1855 as '*Omnis cellula e cellula*' ('all cells (come) from cells'). Virchow was partway through a staggeringly successful career (he had already discovered leukaemia and would go on to essentially invent public health) but nevertheless opted not to give Remak any credit, presenting the ideas as his own. Three years later, he begrudgingly acknowledged Remak's contributions, but only partially.

This chain of discoveries (and one act of flagrant plagiarism) means we have an answer to our question. If all life is made of cells, then presumably the first living organism was also a cell – albeit one simpler than modern cells. The question: 'How did life begin?' thus becomes: 'How did the first cell form?'

Some readers may cry foul at this point. Surely viruses are living things even simpler than cells? After all, they are far smaller than cells, so much so that they were not discovered until the 1890s – over 200 years after Leeuwenhoek spotted his animalcules. A typical virus is nothing more than a simple shell protecting a few genes. However, this simplicity is also why viruses are not good candidates for the first organism. They are so stripped-down that they cannot survive on their own. To reproduce and flourish, viruses must infect cells and take over their internal machinery. Arguably they are not really alive. They may well have evolved later in the history of life, once there were cells for them to parasitise. Cells, it seems, are the only candidates for the first life.

Let's now consider where life might have formed. This question, at least, is fairly easily answered. The only place we know of where life exists, or has ever existed, is Earth. Therefore, it seems reasonable to suppose that our planet was also the site of life's birth. A handful of scientists do argue otherwise, instead pushing the notion that life formed somewhere else in the universe and was then transported to Earth. There are several variants on this 'panspermia' idea and we'll come back to them in more detail in chapter 6. For now, it should suffice to say that we haven't found life anywhere else we have looked, which rather cuts the whole idea down at the knees.

Finally, there is the question of timing. How long was the window of opportunity for life to form on Earth? To answer that, we need to know two things: how old is the Earth, and how long has it had life – which in practice means, what is the oldest known evidence of life?

Of these two problems, the age of the Earth was solved first.

It's traditional, when discussing this question, to begin by making fun of James Ussher, an Irish archbishop who in 1650 claimed that the Earth was created by God at nightfall on 22 October 4004 BC.[18] Ussher reached this conclusion by adding up the ages of key figures in the Bible and cross-referencing with what was then known of ancient history and astronomy. Clearly, his date was wildly wrong, but it's rather unfair to mock him. Ussher was writing before modern science had really got started.

In 1650 Copernicus and Galileo had been and gone, but Isaac Newton was a mere eight-year-old and the entire science of geology lay far in the future. Against that background, Ussher did sterling work. The palaeontologist Stephen Jay Gould defended him stoutly, pointing out the sheer complexity of deducing a reliable timescale when the Bible so often fails to give dates, necessitating finding links with Roman and Persian history.[19] 'Ussher represented the best of scholarship in his time,' Gould wrote. His one mistake was to assume, as many did at the time, that the Bible was a wholly reliable source.

Over the next two centuries, scientists realised that the rocks of Earth's crust form distinct layers called strata, and that the deeper strata are older than the shallower ones. Each stratum represents a period of Earth's history, and these were ultimately given names, such as 'Jurassic'. By the mid-1800s, geologists like Charles Lyell had realised that each stratum had been laid down, grain by grain, over thousands upon thousands of years – but there seemed to be no way to put a reliable figure on it.

One of the first rigorous calculations was performed by the nineteenth-century physicist William Thomson, who was ultimately ennobled as the first Baron Kelvin. Thomson was a devout Christian who was sceptical of Darwin's theory of evolution. He was also an expert on thermodynamics: the science of heat, and in particular of how rapidly objects cool. Kelvin started from the assumption that the Earth was initially very hot, because it gets warmer the deeper into the planet you go, implying that the Earth is losing heat. He estimated in 1864 that it was between 20 million and 400 million years old.[20] It could not be any older and still be as warm as it is. By 1897 he had settled on 20 million years.

Kelvin was also hopelessly wrong, but again it was not really his fault – not entirely, anyway. One pivotal discovery, which began in 1896, was radioactivity. Deep inside the Earth there are plenty of radioactive rocks, and these are a source of heat that Kelvin knew nothing of. He also believed that the Sun was similarly young, because at the time nobody could understand how a star might shine for untold millions of years. It was not until the discovery of nuclear fusion in the 1930s that it was realised that the Sun has a staggeringly huge power source, and could be truly ancient – in which case the Earth could be too.[21]

The discovery of radioactivity would prove to be the key to pinning down the age of the Earth. It is a story in its own right, but we'll skip to the end. In the early twentieth century, physicists realised that some atoms are unstable and tend to break down or 'decay' into smaller, more stable atoms – releasing tiny packets of radiation in the process.

This happens because an atom is not, as was long imagined, a single particle, but instead is made up of three kinds of smaller particles. Each atom has a central core or 'nucleus' made up of differing numbers of protons and neutrons, which is surrounded by a cloud of electrons. The nucleus is the crucial bit, because the protons and neutrons have to be packed together in certain ways. If there are too many or too few of one of them, the nucleus becomes wobbly.

Each radioactive element decays at a predictable rate. Imagine you had a block of uranium-238, the most common form of uranium, which contained 1000 atoms. It would take 4.468 billion years for half (500) of those atoms to decay into lead atoms. It would then take another 4.468 billion years for half (250) of the remaining uranium atoms to decay – and so on until there were none left and the entire block was made of lead. That 4.468-billion-year halving time is uranium-238's 'half-life'. Because every radioactive element has its own half-life, which can be determined by experiment and does not change, they can be used to calculate the ages of rocks. The first person to try this was an American radiochemist called Bertram Boltwood, who in 1907 measured the relative quantities of uranium and lead in rocks and concluded that they were at least 400 million years old.[22]

However, it quickly became apparent that the problem was more complicated than it had seemed, because uranium was not the only radioactive element that decayed into lead. There was also more than one kind of uranium, and they decayed at different rates.

The person who solved the problem, largely single-handedly, was a 'quiet, unassuming' British geologist named Arthur Holmes.[23] He published his first paper in 1911, having only graduated in 1909. In it, Holmes successfully dated a rock from the Devonian period: a time when the first land plants were flourishing and fish first came to dominate the oceans. He concluded that the rock was 370 million years old – a figure that still fits the Devonian as we understand it today.[24]

Two years later Holmes produced his first book, *The Age of the Earth* – despite having in the meantime gone mineral-prospecting in Mozambique, a trip that led to a near-fatal case of malaria. In the book, he made the case that radioactive decay could be used as a reliable clock to determine the age of the Earth. Based on the rocks that had been dated thus far, he stated that the Earth was probably 1600 million years old.[25]

Holmes kept plugging away at his radiometric dating for the

next two decades. He revised his book twice, in 1927 and 1937, by which time he had found rocks 3 billion years old – a figure he was still pushing in 1946.[26]

At this point the story gets complicated, because after years spent making the case for his methods Holmes suddenly found that radiometric dating was widely accepted, and other researchers piled in. The dating techniques were refined, and researchers also began using samples from meteorites – which presumably formed at the same time as the Earth, but have not been subjected to the endless weathering and general brouhaha of the planet.

The crucial year was 1953, when two researchers independently arrived at basically the correct figure. The first into print was Fritz Houtermans, a German researcher who survived being imprisoned by both the Soviet Union and the Nazis, and went on to become a world expert on radiochemistry. He examined the lead content of a meteorite and estimated that the rock was 4.5 billion years old – and argued that the Earth must be the same age.[27] A few months before, at a conference, Clair Patterson had presented similar data from the Canyon Diablo meteorite, which produced the enormous Barringer Crater in Arizona.[28] Patterson generally gets the credit for being first; a frankly ridiculous distinction, when the two studies came out so close together and credit could easily be shared. Patterson subsequently cleaned up some problems with his results and published them properly in 1956.[29] By that point he had settled on a figure of 4.55 billion years for the age of the Earth.

And there, pretty much, it has stayed.[30] More recent estimates have revised it down ever so slightly to 4.54 billion years, and presumably the number will be fiddled with some more, but at this point it's doubtful that it will change substantially. Our planet is a smidge over 4.5 billion years old. Before that time, the Sun was being orbited by a collection of boulders and dust. After that time, there was a young world.

That figure of 4.5 billion years ago is the upper boundary for the age of life on Earth. It seems highly unlikely that life formed any sooner, or at least if it did none of it survived. That's because soon after the Earth formed, it appears to have been hit by a rock about the size of Mars.[31] The entire surface of the planet was melted and huge volumes of rock thrown into orbit – ultimately forming the Moon. Anyone who wants to argue that life is older than 4.5 billion years needs to present a compelling story for how it might have survived such an apocalyptic event. But it is simpler to assume that life began later.

Quite how much later is uncertain. Palaeontologists keep pushing back further into the deep past, finding ever-older fossils and other traces of life, so the window is narrowing.

In the nineteenth century, the fossil record only went back as far as the Cambrian period, which began 541 million years ago. Cambrian rocks hold a rich diversity of fossils, including the woodlouse-like trilobites, worms and sponges. However, many modern groups are completely missing: there were no Cambrian mammals, birds or insects. All known life was confined to the sea.

When the first palaeontologists dug down into older rocks, they found nothing. There seemed to be no 'Precambrian' fossils at all. This proved to be an enormous thorn in Darwin's side as he tried to convince people that evolution could explain life's diversity, because to all intents and purposes it seemed that an enormously rich marine ecosystem had popped up out of nothing.

Not until 1957 was the fossil record extended deeper into the past. In that year, a schoolboy named Roger Mason and his friends went climbing in Charnwood Forest in Leicestershire, England. Mason found a fossil that looked like a fern. He took a rubbing of it and later showed it to his father, who in turn showed it to a local geologist named Trevor Ford, who wrote up the fossil the following year.[32] Coming from undisputed Precambrian rocks laid down tens of millions of years before the Cambrian, this was the first Precambrian fossil to be generally accepted – others had been found over the previous few decades, but were incorrectly lumped into the Cambrian.[33] The fossil was named *Charnia masoni*, after the forest and its young discoverer. It would be a delightful story, except that a girl named Tina Negus had spotted the fossil a year before Mason, only to be roundly disbelieved. It is hard not to see sexism at work.

Since then, the fossil record has been pushed far further back. The oldest traces of life that are entirely uncontroversial are found in the Pilbara region of Western Australia. First described in 1980, this ancient ecosystem is preserved as layered fossils called stromatolites.[34] These were once thin mats of bacteria, which became covered by sediment – at which point a new layer of bacteria formed on top, and so on. They have been studied ever since, and there is no meaningful doubt of their age: they are 3.5 billion years old.

In theory, that gives us a window of a billion years in which life could have formed. But in practice the window is smaller than that. The Australian microbes had complicated internal structures and could link together into chains. They do not look like the first, simplest life, but rather like an intricate, modern-looking bacterial ecosystem. The implication is that life is rather older than 3.5 billion years.

The exact length of the time window is still being hashed out. For instance, until recently it was thought that the Earth suffered a particularly heavy battering from meteorites between 4 billion and 3.8 billion years ago. This 'Late Heavy Bombardment' was thought to have rendered the planet hostile to life, implying that life began no earlier than 3.8 billion years ago. However, simulations suggest that at least some life could have survived.[35] Besides, the evidence for the Late Heavy Bombardment is not as solid as it once seemed. There appears not to have been one short sustained blast of impacts at all.[36] Instead, large meteorites hit from time to time until about 3 billion years ago.[37] So it seems 3.8 billion years ago is not a hard barrier.

There is no shortage of claims to have found older traces of life. For instance, in 2017 researchers described tubes and strands in rocks in Quebec, Canada.[38] They look like microorganisms and the surrounding rocks have chemical traces that suggest the presence of life. The entire assemblage is at least 3.77 and possibly 4.28 billion years old. However, many other researchers are not convinced that the tubes and strands are really microorganisms, as they could simply be unusual rocks. This is a recurring issue with fossils of this age.

Instead of looking for fossils, other teams have looked for purely chemical evidence of life. A study published in 2015 found fragments of carbon preserved in a crystal that formed 4.1 billion years ago.[39] The carbon seems to have been affected by living organisms. There are several forms of carbon, known as isotopes, which differ only in the number of neutrons in the nucleus of each atom. Living things prefer to use carbon-12, so living matter tends to have lots of carbon-12 and little of the heavier carbon-13. That is exactly what was found in the preserved carbon, suggesting it came from living organisms.

Results like these are still being actively discussed, so nobody is really sure when life formed. But it is definitely more than 3.5 billion years old, and it would be foolish to bet against it being over 4 billion years old. This means the window for life's origin is narrow: a billion years at most, and possibly just half a billion or even less. Therefore, whatever explanation we come up with for life's beginning, it cannot rely too heavily on good luck. If Earth had been lifeless for billions of years, it might have been reasonable to postulate something staggeringly unlikely, because there was plenty of time for implausible things to happen. But it may even be that life formed almost as soon as the Earth's surface solidified after the Moon-forming impact. The implication is that the genesis of life is something that can happen relatively easily, and therefore repeatably.

That is where things stand. We have good reason to think that life began on Earth, and nowhere else. It did so after the planet formed 4.5 billion years ago, possibly rather quickly as there may well have been life 4.1 billion years ago. Finally, either the first life was a cell or it swiftly gave rise to one, because we know of no other form of life. So we can now spell out the problem of life's origin in a more specific way: in no more than a billion years, lifeless chemicals must have assembled themselves into a living cell. The question is, how?

Chapter 2

A Soviet Free Thinker

The Russian scientist Alexander Ivanovich Oparin was the first person to publish a hypothesis of life's origin that other scientists took seriously. He imagined simple blobs of fatty material floating in the primeval seas, which ultimately became more complex and gave rise to living cells. This idea opened up the question of life's origin for other scientists, by giving them a working hypothesis to consider. Ultimately, it would help to inspire an experimental breakthrough.

However, as well as contributing his ground-breaking ideas, Oparin was also a profoundly contradictory, controversial figure, because he worked in the Soviet Union under Stalin. Arguably, Oparin was complicit in many of the Soviet regime's abuses. Yet it can also be argued that he had little choice, for resisting would have meant risking his job and possibly his life. Either way, he cannot be fully understood without grasping the historical context.

Oparin came into the world when the Romanov dynasty that had ruled Russia for three centuries was in decline and heading for its catastrophic fall. He was born on 2 March 1894, to a merchant family in the small town of Uglich in western Russia. In the November of that year, the last tsar, Nicholas II, came to power following the death of his father Alexander III. Badly unprepared for the throne, Nicholas led his country into a hopeless war with Japan in 1904, and only barely survived an attempted revolution the following year. A decade later he took Russia into the First World War, despite its army being grossly underprepared, and again the country suffered enormous losses. With Russia on the verge of collapse, the stage was set for the revolution of 1917.

In March of that year, as Oparin turned twenty-three, the Romanov dynasty fell. Entire army regiments mutinied and Nicholas, unable to hold onto his throne, abdicated. A provisional government came to power, but in October it was deposed by the Bolsheviks, led by Vladimir Lenin. The October Revolution would transform Russia into the Union of Soviet Socialist Republics (USSR), a communist dictatorship run according to the doctrines of Karl Marx. The government took control of almost all aspects of the economy and life, in the name of empowering the working people of the country. It hardly needs saying that this did not work out. Instead of becoming a socialist idyll of communal living, the nation became a savage totalitarian dictatorship dominated by a cruel secret police.

On 17 July 1918, Nicholas and his wife and children were taken into the basement of the building where they were staying under house arrest. There they were gunned down and bayonetted by Bolshevik soldiers, and their bodies thrown into an unmarked grave.

This brutal killing was a forerunner of things to come. First under Lenin and then his successor Joseph Stalin, the Soviet regime imprisoned, tortured and killed anyone who presumed to disagree with it. Its Gulag system of forced-labour camps has become a grim legend for us all of what can happen when a government wields too much power, and when a leader builds up a cult of personality that cannot tolerate criticism.

The Soviet state's persecution complex extended far beyond its obvious political opponents. Its doctrines ultimately seeped into every aspect of life, including science. While the government prided itself on being 'scientific', it disallowed scientific findings that seemed to contradict its ideology.[1]

Alexander Oparin would ultimately have to face the Soviet state's oppression of science head-on. However, when he first began thinking about life's origin, the worst of this was years in the future. Oparin had grown up to be a tubby, jolly fellow, fond of smart bow-ties and suits that belied his modest origins. He put people in mind of Count Ilya Rostov in Tolstoy's *War and Peace*: full of bonhomie and zest, determined not to let the darkness in, and fluently quoting classic authors like Pushkin.

Oparin first set out his ideas about the origin of life in a booklet published in Russian, simply called *Origin of Life*.[2] It appeared in 1924: the year Lenin died and Stalin rose to power. To read it, even now, is to discover a minor masterpiece of ingenious hypothesising and deft scholarship.

Oparin began by asking what divided living things from non-living things. He identified three things that seem unique to life: a specific structure, an ability to take in energy and reproduce, and an ability to respond to stimulation. But he then argued that none of these things is truly unique to life. For example, living organisms certainly have a complex internal structure: even single-celled organisms like bacteria are not simple bags of formless jelly, but are 'extraordinarily complicated'. However, Oparin pointed out that many non-living things also form startlingly complex structures. He encouraged readers to think of the 'ice flowers' that form on windows on cold days. 'In their delicacy, complexity, beauty and variety these "ice flowers" may even look like tropical vegetation while all the time being nothing at all but water, the simplest compound we know,' Oparin wrote.

In other words, we should not be fooled into seeing life as being more different from non-living matter than it really is. 'We have no reason to think of life as being something which is completely different in principle from the rest of the world,' Oparin wrote. 'The specific peculiarity of living organisms is only that in them there have been collected and integrated an extremely complicated combination of a large number of properties and characteristics which are present in isolation in various dead, inorganic bodies. Life is not characterized by any special properties but by a definite, specific combination of these properties.'

With that in mind, Oparin began to envision what might have happened when the Earth was young. He imagined the newly formed planet as a seething-hot ball that slowly cooled. Hot steam in the atmosphere mixed and reacted with carbon compounds on the surface, forming simple organic compounds 'capable of further transformation'. At the same time, simple nitrogen compounds like ammonia probably formed. The young Earth had become a natural chemical factory, forming ever more complex and varied organic compounds.

Now, Oparin reasoned, something dramatic happened. 'Finally the time came when the temperature of the surface layers of the Earth fell to 100°C.' That meant water could exist as a liquid instead of a gas. 'Continuous downpours of rain fell upon the surface of the Earth from the moist atmosphere. They inundated it and formed a cover of water in the form of the original boiling ocean. The first organic substances which had hitherto remained in the atmosphere were now dissolved in the water and fell to the ground with it.' In this primeval ocean, 'ever larger and more complicated particles were formed'. Among them might even have been carbohydrates and proteins, 'the foundation of life'.

As ever more complicated molecules formed, eventually some of them started to form blobs of jelly-like substances. Such mixtures are called 'colloids' and form naturally when long, stringy molecules mix with water. It marked another crucial transition, because these jelly-like blobs looked a bit like independent cells, and their interiors were separated from the surrounding water. These primeval cells had the beginnings of complex structure, and they were also distinct individuals. As Oparin put it, 'we can even consider that first piece of organic slime which came into being on the Earth as being the first organism'.

Over the years many such slime blobs formed, each with a different mix of chemicals. The action of ocean waves broke some of them apart, in an early form of cell division. Separate 'species' were formed, each with its own chemical makeup. Now natural selection could take effect, favouring the blobs that were better at obtaining crucial chemicals and sustaining themselves. Over many generations and thousands of years, the primordial cells got better at feeding themselves by guzzling the chemicals in the sea.

That is the essence of Oparin's idea: the first step-by-step process by which non-living chemicals could assemble themselves into a simple cell. He argued that there was an unbroken chain of descent from the first simple blobs of jelly in the primeval ocean to modern living cells.

He was evidently aware of the obvious objection to his idea: that nobody has ever observed such a simple cell. 'It is true that no trace of these primitive living things now remains on the Earth, but this is no proof that they never existed,' he argued. 'It should not be forgotten that at a certain period of the existence of the Earth they must have been completely wiped out by their more highly organized comrades.'

Despite its brilliance, Oparin's little book was almost entirely ignored, especially outside Russia where it remained unknown.[3]

Five years later, in 1929, the British biologist John Burdon Sanderson 'J. B. S.' Haldane published a substantially similar idea. The two men worked entirely independently and there is nothing to suggest Haldane was aware of Oparin's work. In temperament they seem to have been almost entirely different, but both were communists. This seems to have enabled their thinking in this direction, because communism was built on a philosophical foundation called dialectical materialism, which entailed explaining everything in terms of material objects. So while most of Western society was still half in thrall to Christian ideas, communists like Haldane and Oparin were vigorously rejecting them and seeking physical explanations for even 'spiritual' phenomena like life and the soul.

Haldane was larger-than-life: so much so that Ronald Clark's biography *J. B. S.* is essentially one long stream of outrageous anecdotes punctuated by occasional outbreaks of science.[4] He fought in the First World War, a bloodthirsty business that he later admitted he had enjoyed – a response that troubled him ever after. In later life he embraced a philosophy of non-violence, which extended even towards insects, partly inspired by Hindu teachings. Haldane was drawn to communism because he was a humanitarian convinced that society was unfair and cruel. It's difficult not to admire the ornery old malcontent for his convictions, even though they sometimes led him to be overtly sympathetic to men like Stalin, who were the opposite of humanitarian.

After the war ended, Haldane immersed himself in academia, first at the University of Oxford and then at the University of Cambridge. His research ranged widely, but his most significant achievement was to apply his mathematical talents to the theory of evolution, discovering ways to unify it with the new science of genetics.

By this time, geneticists had shown convincingly that inherited characteristics like height are carried by things called genes. Nobody knew what genes were made of or how they worked, but it was possible to trace their passage through the generations. The Austrian monk Gregor Mendel had hinted at this by studying pea plants, which unlike humans are either tall or short; there are no medium-height pea plants. By meticulously crossing one pea plant with another, Mendel showed that each plant carried two unidentified substances (the word 'gene' would not be coined for many years) that together controlled its height. If either of them was the 'tall' version, the plant would be tall. Only if both were 'short' would the plant be short.

Between 1924 and 1934, Haldane produced ten papers. According to Clark, 'he was one of the three men whose evidence showed not only that Darwin's theory of evolution could work but exactly how it could work. What he did was to produce a formula which gave a numerical indication of how Mendelian genetics would work if Darwin's theory of natural selection was correct.' This was a crucial contribution to what became known as the 'modern synthesis', a deeper version of Darwin's theory that incorporates genetics and shaped our understanding of evolution for decades.

In 1924, when Oparin published his book, Haldane was also doing big things. As well as publishing the first of his papers linking evolution and genetics, he met his first wife Charlotte Franken, a journalist and feminist. Charlotte was already married, necessitating a tricky divorce that caused some scandal and nearly got Haldane fired from Cambridge, but the two ultimately married in 1926.

J. B. S. and Charlotte established themselves in Cambridge, and an incident that took place shortly afterwards illustrates his character. One of Haldane's students, Martin Case, was charged with dangerous driving. The chief prosecution witness was a night watchman, so Haldane devised a plan to stop him testifying. The man was known to frequent a particular pub, so Haldane headed over at opening time on the day of the trial and got into conversation with him. He then 'plied his victim relentlessly with booze for nearly three hours', until the man was virtually incapable of speech. The ensuing trial was a shambles: the star witness could barely supply his own name, and his account of the actual incident was confined to repeatedly blurting out 'Ecumuptheillikefuckinellanwentarseovertip', belching, and finally collapsing onto the floor of the witness box. The case was duly dismissed, and Case's profuse gratitude to Haldane was only slightly tempered when he was presented with an itemised bar bill of 'staggering' magnitude.

The incident is revealing. Haldane was deeply loyal and was willing to stick his neck out for people he believed in, but there was also a recklessness about him. There doesn't seem to be any doubt that Case really had been driving dangerously, but

Haldane evidently didn't care too much about that.

In 1929, Haldane published a short article in the *Rationalist Annual* in which he outlined a hypothesis for the origin of life that was in many ways remarkably similar to Oparin's.[5] It was not a scientific paper, but rather a piece of gently speculative popular-science writing.

Haldane began, as Oparin did, by imagining the new Earth cooling from its molten state until oceans could form, whereupon a mix of carbon- and nitrogen-based compounds were formed, all bathed in ultraviolet radiation streaming from the Sun. 'Now, when ultra-violet light acts on a mixture of water, carbon dioxide, and ammonia, a vast variety of organic substances are made, including sugars and apparently some of the materials from which proteins are built up,' he wrote. 'In this present world, such substances, if left about, decay – that is to say, they are destroyed by micro-organisms. But before the origin of life they must have accumulated till the primitive oceans reached the consistency of hot dilute soup.'

In that last sentence, Haldane achieved something that Oparin did not: a catchy phrase. For that mention of 'hot dilute soup' is the origin of the term 'primordial soup'. When we speak of life emerging from the primordial soup, we are unknowingly quoting Haldane.*

* Alternatively, the phrase may come from John Butler Burke, whose origin-of-life experiments involved bouillon broth. There is also the variant form 'primeval ooze', which seems to date back further. For instance, in his 1870 book *Die Zeugung* ('Generation'), German naturalist Lorenz Oken argued that all life began with the *Urschleim*: 'primordial slime'. It is hard to find a hypothesis for the origin of life that doesn't somehow revolve around a sticky, gloopy liquid.

At this point, Haldane diverged from Oparin. Whereas Oparin imagined that the next step was the formation of blobs of jelly, the precursors of cells, Haldane supposed that the next phase was instead molecules that could make copies of themselves. 'The first living or half-living things were probably large molecules synthetized under the influence of the Sun's radiation, and only capable of reproduction in the favourable medium in which they originated. Each presumably required a variety of highly specialized molecules before it could reproduce itself, and it depended on chance for a supply of them.'

Cells, which acted as containers for the replicating molecules, would come millions of years later. 'The cell consists of numerous half-living chemical molecules suspended in water and enclosed in an oily film,' Haldane reasoned. 'There must have been many failures, but the first successful cell had plenty of food, and an immense advantage over its competitors.'

Haldane's idea is much sketchier than Oparin's, largely because he had fewer words to play with. Nevertheless, the central image of the primordial soup would stick. Once Oparin's work became more widely known, the idea would be labelled the Oparin–Haldane hypothesis.

This rather obscures the fact that the two men's ideas were different in crucial respects. While Oparin emphasised the importance of cells and nutrition, Haldane's focus on self-replicating molecules implies – although he didn't explicitly state this – that life began when genes arose. At the time this seemed like a subtle distinction. It would not remain so.

Over the following years, as Haldane busied himself with his numerous research interests and with campaigning for communism, Oparin fleshed out his ideas about life's beginning.

In 1936 he published a full-length book, *The Origin of Life on Earth*.[6] This time its audience would not be limited to Russian scientists, because two years later it was translated into English and published widely. This was the book that would take Oparin's ideas to the audience they deserved. Ironically, it is much harder to read than his earlier work. It is full of tedious detail: two full pages are spent quoting Engels on materialism, and there are over 100 pages of preamble before he finally gets down to business.

The one significant advance came when Oparin discussed the first primitive cells. Whereas before he spoke simply of 'jelly-like' substances, he now had something more specific in mind: coacervates (pronounced 'coh-AH-ser-vates').[7] These are made when mixtures of long chain-like molecules called polymers, floating in water, experience a sudden change like a rise in temperature. Instead of intermingling with the water molecules, the polymers gather into spherical blobs called coacervates. Each droplet surrounds itself with a semi-rigid 'wall' of water molecules. These are precisely aligned and keep it separated from the surrounding water.

Coacervates are remarkably lifelike. For starters, they maintain their integrity. 'These droplets may fuse with each other but they never mix with the equilibrium liquid,' Oparin wrote. What's more, they can grow by taking in more polymers, and sometimes divide into two rather like true cells.

Oparin was not arguing that coacervates are the same thing as the 'protoplasm' found inside cells, which he acknowledged was surely more complicated. But he thought one might have led to the other.

Oparin also gave a more detailed discussion of how these first coacervate-based 'cells' might have become better able to sustain themselves. He focused on catalysis: the ability of certain substances to speed up chemical reactions, without themselves being permanently affected. Often, two chemicals react only slowly with each other if they are alone, but react much faster if even a tiny amount of a catalyst is added. Living organisms make their own catalysts in the form of complex molecules called enzymes, which often work extraordinarily well. Enzymes are absolutely crucial to many of the things living organisms do. 'In the last analysis, all vital phenomena such as nutrition, respiration, growth, etc., result from the chemical transformation of organic substances,' Oparin wrote. Even the simplest living cells carry out a great many different reactions. Staying alive requires 'a number of chemical reactions following one another in a perfectly definite order', says Oparin.

How might such an intricate system have arisen? Oparin supposed that each primordial cell held its own mix of chemicals, which reacted slowly with each other. In some cases those reactions would have caused the cell to break up, while in other cases they would have helped preserve it. It was a simple form of evolution by natural selection. Over time, the blobs that were better at sustaining themselves became more common.

This is the end point of the argument that Wöhler inadvertently unleashed a century before when he made urea. For Oparin, all the seemingly mysterious properties of life are, on closer inspection, the result of chemical reactions.

Finally, Oparin stated an idea that would become a founding assumption for studies of life's origin: that it was a slow, tortuous process. Life, he reminded his readers, is 'infinitely more complex than any simple solution of organic substances'. It would be 'senseless' to suppose that something so complicated could originate quickly.

Oparin had made his name with his book, but he was also about to sully it. So was Haldane. Both men were about to come face to face with the evils of the USSR.

In the 1930s, the Soviet state began to interfere in earnest with the workings of science, most dramatically in agriculture and genetics. The central character was a peasant named Trofim Denisovich Lysenko, who has now become a bogeyman figure that scientists use to scare their students with tales of what can happen when science becomes overtly ideological. Lysenko was focused on trying to improve the yields of wheat crops, as Russia had long been prone to severe famines. But his plans were always cock-eyed, concocted in defiance of the evidence. Lysenko's dominance did nothing to prevent more famines.

Lysenko became a favourite of the government partly because he was a peasant, not a member of the despised bourgeois elite. But he also presented his daft ideas in a way that highlighted how well they corresponded with state doctrines. The central point was that Lysenko completely rejected the science of genetics. He declared that genes did not exist and that Western scientists were 'idealists' for studying a fictional construct – and persuaded the Soviet establishment to support him.

It all came back to the wheat plants, which Lysenko had promised to drastically improve. Genetics implied that this would be slow work, because genes only change at random: a process called mutation. This became clear from the 1920s onwards, when geneticist Hermann Muller began exposing fruit flies to low doses of radiation to trigger mutations, and obtained an array of unusual variants.[8] In wild animals, mutations happen relatively rarely. Sometimes they are helpful but more often they are harmful, and natural selection tends to winnow out all but the useful ones. This is how evolution works, and it is a slow grind. A person might work all their life to build up their muscles, or to improve their intelligence, but none of that improvement will be passed on to their children.

Instead, Lysenko asserted that organisms could be transformed by changing their environment. Instead of near-fixed genes that had to be shaped by a slow process of evolution, it was possible to radically change organisms through a kind of shock treatment, and these changes would then be passed on to their offspring. It is easy to see how appealing this was in the USSR, where Stalin and his allies were trying to reshape society and even human nature itself.[9]

In this context, it is striking that Oparin's entire opus barely mentioned the word 'gene', and his hypothesis made no attempt to explain the origin of genes. Instead, he focused on creating the shell of a cell (in the form of a coacervate) and on how it might have fed itself. Enzymes were his speciality, so it is natural that he concentrated more on them, but it does not feel like a coincidence that he so entirely ignored the forbidden science of genetics. Speaking of genes in the 1920s would have been safe enough, but by the 1930s Lysenko was ascendant and it could have been dangerous for Oparin.

Haldane, living in Britain, was seemingly insulated from all this – but in fact he was enmeshed in it. He had visited Russia in 1927 with his wife Charlotte, and the experience hardened his left-wing views and his admiration of the Soviet state. He had been invited by Nikolai Vavilov, an expert in plant genetics, who was admired in Russia for his work on improving crops like wheat. But by the 1930s, Lysenko had secured a great deal of power for himself and was remorselessly attacking 'bourgeois' genetics. The situation for Vavilov and other Soviet geneticists became precarious: they faced public condemnations and threats.

However, Haldane, like many communist sympathisers in the West, was too slow to recognise what was happening and downplayed the alarming events. In July 1939 Vavilov unexpectedly announced that he could not take part in a major genetics conference in Edinburgh, and sent two conflicting explanations in separate letters. Haldane, aware of this, must have realised Vavilov was acting under duress. But two months later war broke out in Europe, and any discussion of Soviet science became conflated with arguments about whether they would be Britain's allies in the conflict.

In August 1940, Nikolai Vavilov was arrested. The following year he was sentenced to death. This was commuted to twenty years' imprisonment, but in 1943 he starved to death in prison – a grimly ironic end for a person who worked all his life to prevent food shortages. News of his death would not reach the West for some time, and Haldane was still giving Lysenko the benefit of the doubt in print in 1945.

The Lysenko affair came to a crisis in 1948, and now it was Oparin's turn to behave poorly. The USSR had suffered yet another famine in 1946–7, and on New Year's Eve 1946 Stalin summoned Lysenko to the Kremlin. It got Lysenko his one and only photo opportunity with the great leader, reinforcing his position. It also set him off on another futile scheme to save Soviet agriculture. This time, Lysenko was pushing branching wheat, which produces a photogenically large number of seeds per plant but is almost entirely useless for feeding people because few plants grow per acre.

More knowledgeable biologists decided to have it out with Lysenko once and for all. In early 1948 he was attacked in a memo to Stalin and in a speech in Moscow. Stalin was enraged. He reportedly paced back and forth in his office, repeating, 'How did anyone dare insult Comrade Lysenko?' His response was to hold a sort of show trial, a committee meeting in which Lysenko would defend himself against his critics. Stalin helped write Lysenko's report to the committee and his speech, and dictated a crucial paragraph in which Lysenko announced that 'the Central Committee of the Party had examined my report and approved it'.

This line was crucial. The government had made it crystal clear, from the highest level, that genetics was forbidden. Over the next few days, geneticists began publishing apologetic letters in the Communist Party's official newspaper *Pravda*, abandoning the science of genetics and promising to abide by Stalin's diktat. One can't really blame them: anyone who kept resisting after Stalin had weighed in was liable to end up dead like Vavilov.

However, Oparin went further by actively supporting Stalin's oppression. He published a full-page letter in *Pravda* complaining about 'fenced-in pontiffs toying with fruit-flies' and called for genetics to be renounced, which it soon was. As a result, Oparin soon became head of biological sciences. Geneticists saw their stocks of fruit flies destroyed and their textbooks rewritten, and many were sacked. Things only became more ridiculous in 1950, when Oparin invited the crank biologist Olga Lepeshinskaya to receive a Stalin Prize – despite the fact that she supported the discredited ideas of vitalism and spontaneous generation, rejected the idea that cells only form from other cells, and faked her experimental results.

Can Oparin's actions be excused? Interviewed in 1971, Oparin defended himself by highlighting his terror of the Soviet authorities.[10] 'If you had been here in those years, would you have had the courage to speak out and be imprisoned in Siberia?' he asked. One has to have at least some sympathy: after all, the true horror of totalitarianism is that it makes collaborators its victims. It is also worth noting that Oparin protected younger researchers at the Institute of Biochemistry, which he headed, including the geneticist Andrei Belozersky. Still, Oparin did not simply survive Lysenkoism: he used it to advance his career.

In contrast, Haldane was increasingly disturbed by reports from the USSR. The events of 1948 had put him in the position of defending the indefensible, and he was not willing to bend his belief in the evidence for Mendelian genetics, even to fit his communist ideals. In 1949 he resigned from the Communist Party. While he remained a Marxist for the rest of his life, he would no longer blindly support the Soviet government.

After Stalin died in 1953, Lysenko's influence on Soviet science waned. By the 1960s scientists felt able to criticise him in public, and in short order he lost his position, his hapless methods were exposed, and he was publicly disgraced. However, Oparin managed to survive this tumult, despite his previous association with Lysenko. In 1969 he was awarded the Hero of Socialist Labour, one of the Soviet state's highest honours.

Despite all the Soviet messiness, between them Oparin and Haldane had established a working hypothesis for how life got started. The primordial soup hypothesis gradually gained widespread acceptance: 'so much so,' Clark notes, 'Haldane himself later commented, that his mistrust of orthodoxy made him doubt whether it could be correct'. Yet despite their similar ideas and outlooks, the two men would only meet once, when Haldane's life was drawing to a close.

In October 1963 they both attended a conference on the origin of life in Wakulla Springs, Florida, organised by Sidney Fox (who we'll meet in chapter 7).[11] Oparin spoke little English, so had to communicate through an interpreter. Haldane introduced him as the opening speaker, and cheerfully conceded that the Russian had published the idea first. 'The question of priority doesn't arise,' Haldane said drily. 'The question of plagiarism might.'

It was a fleeting encounter, not to be repeated. Haldane was ill, suffering from rectal bleeding that gave him his first hint of the cancer that would kill him the following year – although not before he wrote a blackly funny poem about his cancer* and recorded his own televised obituary.[12] Oparin's career was also winding down. But by then they both knew that they had been, on some level, successful. For a full decade before their meeting, an American chemist had published an experiment that seemed to confirm the truth of the primordial soup hypothesis.

* A short sample will have to do:
'My rectum is a serious loss to me,
But I've a very neat colostomy,
And hope, as soon as I am able,
To make it keep a fixed time-table.'

Chapter 3

Creation in a Test Tube

The most famous experiment ever conducted into the origins of life, one that would define the entire field for a generation, was an improvised bodge job. It was performed by a young man at the start of his scientific career, with everything to prove, who was fortunate enough to enter the orbit of an older colleague with absolutely nothing to prove.

By the time the experiment came to be published, Harold Urey was sixty years old. He had won the Nobel Prize nineteen years before. He could have rested on his laurels, but he decided to take one last big gamble on an ambitious graduate student: a gamble that would pay off spectacularly.

Urey was born in the small town of Walkerton, Indiana, in 1893.[1] His father died when he was little. The young Harold was raised in a fundamentalist Protestant sect, but lost his faith as a teenager. After graduating from high school he became a teacher, but then enrolled at the University of Montana where he majored in biology and chemistry. Urey was single-minded: he achieved straight As, except in athletics, while also waiting tables and even working a summer on a local railroad. By the time he graduated in 1917, the United States was entering the First World War, and Urey went to work at a local chemical plant. From here on he was a chemist, not a biologist.

After his PhD, Urey went to Copenhagen in 1923, where he met the leading figures in quantum mechanics: a then-nascent area of physics concerned with the behaviour of the smallest, subatomic particles.[2] Researchers like Werner Heisenberg and Niels Bohr had found that particles like protons and electrons behaved in almost inexplicable ways that seemed to defy common sense.

Urey had realised that if he was going to understand how chemicals behaved, he must first grasp quantum physics, because only in the light of quantum physics do atoms make sense. In particular, Bohr had revised the previous picture of the structure of an atom. Instead of the electrons orbiting the nucleus in a fuzzy cloud, he saw that they could only orbit at certain fixed distances. It was as if there were concentric shells surrounding the nucleus, each able to accommodate a set number of electrons but no more. This would ultimately explain why some atoms react with one another, but not others. The atoms are always 'trying' to fill their outermost shell with electrons, or to empty it if there are only one or two. An atom with a space to fill 'wants' to react with one that has a spare electron, but two atoms with spaces to fill are no good to each other.

Armed with this new knowledge, Urey carved out a niche for himself as a physical chemist through the 1920s. Finally, in 1931, he began the project that would net him a Nobel Prize. In August he learned that there was a second variant or 'isotope' of hydrogen, each atom of which weighed almost exactly twice as much as the standard kind. It would later be understood that a normal hydrogen atom only had a proton in its nucleus, while a 'heavy hydrogen' atom had both a proton and a neutron. Urey decided to isolate a sample of heavy hydrogen, which is now called deuterium.

Urey and two colleagues worked frantically for the remainder of 1931 to achieve this. It is a measure of his obsession that he spent Thanksgiving Day examining the samples that confirmed their success, before trailing home, hours late for dinner, to tell his wife Frieda the news.* History does not record her response, but Urey apparently offered the excuse that 'we have got it made'.[3] The result was duly published on 1 January 1932.[4] For his efforts, Urey was awarded the 1934 Nobel Prize in Chemistry. In typical Nobel fashion, it was only given to Urey, even though there were three authors on the paper. He did, however, give his colleagues a quarter of the prize money apiece.

Over the following decade, Urey became the world expert in separating isotopes. With the outbreak of the Second World War in 1939, Urey was one of many scientists who became alarmed at the prospect of the Nazis building an atomic bomb, and who therefore began studying nuclear fission. This ultimately led him to the Manhattan Project: the United States' effort to build the first nuclear bomb. Urey was in charge of a team that sought to separate uranium-235, the atoms of which could be split, from other uranium isotopes. The project hit a multitude of brick

* This is not the most extreme instance of single-mindedness on Urey's part. On one occasion, he was approached by a student excited to tell him some new results, but Urey was himself eager to describe something new. Eventually the student got a word in edgeways and began his account, at which point Urey seemingly fell asleep. The student paused, so Urey opened his eyes and said, 'Now, as I was saying . . .'

walls until finally, in August 1945, the US succeeded in dropping atomic bombs on the Japanese cities of Hiroshima and Nagasaki.

After the war ended, Urey changed course. He had seen enough isotopes for a lifetime and the savagery of the war had taken a toll on him, even though he was far from the fighting. He campaigned for nuclear power to be brought under civilian control, fearful of what the military would do with it, and spoke in favour of a world government. So outspoken was Urey, and so critical of the US government and military, that he was hauled in front of Joseph McCarthy's notorious House Un-American Activities Committee.

Urey turned his attention to the chemistry of outer space, a field of study that he essentially created single-handedly. Returning somewhat to his original degree in biology, Urey began thinking about how life might have begun and in particular the chemicals required.

In late 1951, now at the University of Chicago, Urey gave a seminar about the origin of the Solar System and conditions on the young Earth. He knew that stars are predominantly hydrogen and that the outer planets are rich in methane – each molecule of which contains a single carbon atom surrounded by four hydrogens, arranged in a tetrahedron. With this in mind, he made an educated guess at the contents of Earth's first atmosphere. Today the air is 78 per cent nitrogen and 21 per cent oxygen, the remaining 1 per cent being trace gases like argon and carbon dioxide.* But Earth's first air was surely very

* It's a grim irony of atmospheric science that carbon dioxide makes up less than 1 per cent of Earth's atmosphere, yet has such an outsized effect on Earth's climate.

different. Oxygen is only released by green plants and their ilk, and there were none, so there cannot have been any free oxygen. Urey also thought there was less nitrogen than today. Instead, he suggested that the atmosphere was dominated by methane and ammonia.

Such an atmosphere would only allow certain kinds of chemistry to occur. All chemical reactions are ultimately the result of electrons moving from one atom to another, or being shared between two atoms. The chemicals Urey listed all tend to give away electrons to other chemicals, and do not pick up electrons.

Chemists call such a mixture a reducing atmosphere. 'Reducing' is a confusing word for something perfectly simple. When something gains an electron, chemists say it has been 'reduced'. The opposite process, losing an electron, is dubbed 'oxidation'. So a reducing atmosphere contains gases that donate electrons, 'reducing' anything they touch. These terms were devised long before anyone knew about electrons. Oxidation originally referred to reactions in which oxygen atoms became attached to something else, and reduction meant the opposite reaction in which oxygen was removed – but the terms were later applied more generally after chemists learned about electrons. Arguably, it would have been better to simply make up new names, but nobody did, so we are stuck with ludicrously misleading designations that take an entire paragraph to unpack.

At any rate, Urey suggested at his seminar that a reducing atmosphere, blasted by lightning strikes and bathed in ultraviolet light from the Sun – for no oxygen meant no ozone layer – would have been an ideal chemical factory for organic compounds.

This is essentially what Oparin and Haldane had suggested when they described the primordial soup, but Urey did not know that at the time. Earlier that year, researchers at the University of California, Berkeley had tried to make organic compounds in an oxidising mixture of carbon dioxide and water, but even after bombarding it with high-energy helium ions they got very little.[5] Urey remarked that 'perhaps a new idea is needed' and that someone ought to try synthesising the chemicals of life in a reducing atmosphere.

Sitting in the audience was a young graduate student named Stanley Miller. Urey's lecture would change the course of his life: it is no exaggeration to say that it was the most important thing that ever happened to him.

Miller was born in Oakland, California, in 1930 – the year before Urey began his furious effort to obtain heavy hydrogen. With a lawyer for a father and a former schoolteacher for a mother, he was an eager reader and a 'chem whiz' at high school. He was shy and enjoyed his own company: he was particularly fond of Boy Scout summer camp because it meant lots of time reading by himself.[6] He also had a lifelong fondness for steam trains, and later in life built a steam car.

After graduating from the University of California, Berkeley, Miller went to the University of Chicago for his PhD. It was one of the few universities that offered paid teaching assistantships, and after his father's death he needed the money. There he heard Urey's lecture – and promptly decided to work on a theoretical project with the nuclear physicist Edward Teller, who had previously advocated building a more powerful nuclear weapon called the hydrogen bomb.* The aim was to figure out how chemical elements were made when the universe was young. But after a year Miller had made no progress and Teller had moved to California, so he decided to change tack and remembered Urey's talk.

Miller approached Urey in September 1952 and suggested trying to synthesise organic compounds in a mixture of reducing gases.[7] Urey was initially cautious, arguing that Miller ought to do an experiment that was likely to succeed, rather than taking such a gamble. Miller himself must have had doubts, because he was clumsy and not the best practical experimenter, which is why he had first tried something theoretical.[8] But he stuck to his guns and eventually Urey agreed to let him try, on the condition that he get results within a year or drop it.

Miller contrived a simple apparatus that would simulate both the ocean and atmosphere. It consisted of two glass flasks, linked by two glass tubes. One flask contained water and could be heated, mimicking the ocean. The other held methane, ammonia and hydrogen, simulating the atmosphere. It also had an electrode inserted to simulate lightning strikes. The university's glassblower made the lot over the course of a week.

The experiment worked first time. Miller began gently heating the water and triggering the electrical sparks, and after two days the water had turned pale yellow and there was a tar-like residue on the 'atmosphere' flask. Clearly, some chemical reactions had occurred. Anxious to find out what had happened, Miller

* Teller was an enthusiastic backer of nuclear weapons and the nuclear arms race, so much so that he is rumoured to have been one of the inspirations for the character of Doctor Strangelove in Stanley Kubrick's eponymous film.

stopped the experiment and analysed the yellow water using chromatography paper. He was delighted to find a single purple spot in a particular location on the paper. He had made glycine, the simplest amino acid. This was a startling success, because amino acids are one of the most important chemicals in living organisms. They are the building blocks from which larger molecules called proteins are made, and no known organism can live without proteins.

Urey was away at the time, talking up the experiment, and only got the good news when he returned. Miller then repeated the experiment, this time for a full week and boiling the water – hoping to drive the reactions further. The water turned first yellow and then brown, while the atmosphere flask was coated with oily scum. The chromatography paper now showed several amino acids, not just one.

At this point Urey decided to publish. He used one of the privileges of being a Nobel laureate and called Howard Meyerhoff, the editor of *Science*, one of the most prestigious scientific journals. Urey wanted the findings out as soon as possible. Meyerhoff promised to publish within six weeks, so Miller wrote a draft and showed it to Urey. Selflessly, Urey insisted that Miller submit it solely under his own name, giving his student all the credit. They submitted the paper in December 1952, but after six weeks nothing had happened. Enraged, Urey had Miller withdraw the paper and submit it to another journal, at which point a panicked Meyerhoff rang, begging him to reconsider. It turned out that a scientist asked to review the paper had not believed the results, and rather than sending comments back had simply put it aside. Meyerhoff straightened things out and the paper was duly published in *Science* on 15 May 1953 –

just eight months after Miller had proposed the experiment.[9]

During the delay, Miller described the results at a seminar at his university's chemistry department. Such seminars were usually given by distinguished visiting professors, not twenty-three-year-old graduate students. When he stepped out to speak he was confronted by a row of famous faces, who hurled questions at him afterwards. But he held his own, and many who were sceptical beforehand came around. At one point someone – possibly nuclear physicist Enrico Fermi – asked whether such chemical reactions could actually have taken place on the young Earth. Urey leapt in: 'If God did not do it this way, then he missed a good bet.'

By then Miller must have known that he had performed an epochal experiment, and when it was finally published the press loved it. 'If their apparatus had been as big as the ocean, and if it had worked for a million years instead of one week, it might have created something like the first living molecule,' claimed *Time* magazine.[10] The *New York Times* used the evocative headline 'Life and a glass Earth'.[11] Overnight, Miller had gone from an unknown graduate student to global fame. He would struggle with this for the rest of his life, as do so many whose defining work is accomplished early on. None of his later experiments would equal that first one. In the years following the experiment, he began to suspect that he was not a good chemist, and even took painting lessons from a cousin – who, after looking at his art, advised him to go straight back to science. He knuckled down, published his graduate thesis and began performing subtle variations on the experiment, identifying more amino acids and varying the mixture of gases to see how it affected the outcome.

Meanwhile, the study of life's beginnings suddenly took off.

In August 1957, the first international conference on the origin of life was held in Moscow, organised by Oparin.[12] Miller was one of many Western scientists who chose to go, despite the risk of being persecuted by the US authorities. He was approached by officials from the Central Intelligence Agency, who wanted details of the Russians' progress.[13] They seem to have been nervous that the USSR might take the lead in biology, as they had in space travel: the Soviets launched the first artificial satellite, Sputnik 1, in October that year.

Indeed, this first explosion of origin-of-life research can only be understood as a product of the time in which it happened. The Space Race and Cold War dominated everything. Almost all US origins research was funded and driven by NASA, which saw it as crucial to the search for extraterrestrial life. There was also a supremely confident tone to much of the research, akin to the 'conquest of space' rhetoric. There was a sense of untrammelled possibility, of expansion beyond the farthest horizon.

Accordingly, a cottage industry in prebiotic (pre-life) synthesis sprang up after Miller's experiment. The most prominent researcher was arguably Joan Oró i Florensa.

Oró was born in 1923 in Lleida in northern Spain.[14] Like Oparin, he grew up amid brewing conflict. A month before his birth, a military officer named Miguel Primo de Rivera had overthrown the Spanish parliament and soon became the country's dictator. Seven years later democracy was restored, but tensions remained and in July 1936, when Oró was twelve, the Spanish Civil War began. It ended three years later when General Francisco Franco became Spain's dictator. He ruled until his death in 1975.

Oró became fascinated by the origin of life during his teens, driven in part by his growing scepticism towards religion. Accordingly, he wanted to study biochemistry, but Spanish universities did not teach the subject. He finally obtained a degree in chemistry in 1947, then returned to his hometown and tried to earn a living as a chemist. However, after two companies had failed, he was forced to work in his father's bakery. He married his wife Francesca Forteza in 1948 and the couple soon had the first of their four children.

Determined to get into biochemistry, Oró decided to move to the United States. He found a place at the Rice Institute in Houston, Texas, and arrived in August 1952, leaving his young family behind. He spent the next three years working on his doctorate, not daring to leave the country in case he was not allowed back in. Eventually, he landed a job at the University of Houston and brought his family over to live with him.

Finally able to explore prebiotic chemistry, in 1960 Oró pulled a rabbit out of a hat.[15] He made adenine, one of the crucial components of DNA.[16] Suddenly a whole new class of biological molecule seemed to have existed on the early Earth. Oró began with hydrogen cyanide, each molecule of which contains one atom each of hydrogen, carbon and nitrogen. He mixed it with ammonium hydroxide to form ammonium cyanide, which he then heated at 90°C for twenty-four hours. Next, Oró filtered out unwanted black tar, then mixed the remainder with hydrochloric acid – the acid found in human stomachs. This yielded a small amount of adenine. However, Oró's setup was a long way from Miller's simulated sea and air: it needed several steps and powerful chemicals.

Two of Oró's chosen chemicals, formaldehyde and hydrogen

cyanide, would go on to be used in hundreds of similar syntheses.[17] Both are associated with death: formaldehyde was long used to embalm corpses, while hydrogen cyanide was used by the Nazis, in the form of Zyklon B, to murder people on an industrial scale. It is uncanny that both are seemingly crucial in life's beginning.

As the 1960s wore on, Oró also worked on the exploration of other planets and studied organic compounds found in meteorites – which implied that the chemicals of life were present in the rocks that formed the Earth.[18] Later in life, after Spanish democracy was restored, Oró returned to his native country. He died of cancer in 2004. Appearing on Spanish television not long before, he was tranquil about his mortality, saying: 'We're only star-dust . . . I'm happy to return to the stars.'[19]

Around the time Oró published his first results, Cyril Ponnamperuma also entered the fray. Like Oró, he was born in 1923, but in the city of Galle in what was then British-controlled Ceylon and is now Sri Lanka.[20] After first studying philosophy, he switched to chemistry in hopes of a steady wage. Eventually, Ponnamperuma emigrated to the United States. In 1962 he began researching exobiology – life on other planets – at NASA's Ames Research Center in California. Like Oró, he made his name by producing crucial organic chemicals, and later became involved in space exploration.[21]

His biggest achievement, published in 1963, was making adenosine triphosphate (ATP).[22] Chemically, this was similar to the adenine that Oró made. Each molecule of ATP includes an adenine that is attached to a sugar called ribose, which in turn carries a chain of three phosphates. ATP was discovered in 1929 and by the 1940s it was clear that it is critical to life.[23] It is a chemical battery: the energy organisms get from food is stored in ATP, and released when needed.

Ponnamperuma had recruited a young scientist named Carl Sagan, who later achieved global fame thanks to his television series *Cosmos*.[24] Sagan was not quite thirty and success was a long way off. He was being divorced by his first wife, microbiologist Lynn Margulis, because he was over-focused on his career and did not help raise their two children – leaving Margulis's own career in limbo.[25] Margulis ultimately became one of the most influential biologists of the twentieth century.*

Sagan had written an undergraduate essay for Urey on the origins of life and was eager to pursue the subject. He suggested that a mixture of adenine, sugars and phosphates might form ATP, if drenched in ultraviolet radiation. A lab technician named Ruth Mariner did most of the experiments, and succeeded in making a small amount of ATP. It was a big success, but it soon came in for criticism: it was not clear that phosphate was present on the early Earth in the quantities needed.

Meanwhile, Miller kept plugging away. By the 1980s he was something of an elder statesman, who could be relied upon to comment on any new idea – often critically. However, his career came to an abrupt end in 1999, when a series of strokes robbed him of the power of speech and left him confined to a nursing home. Jeffrey Bada, who did his PhD with Miller from 1965 to 1968 and worked with him afterwards, often visited him. A therapist was trying to teach Miller to write his ABCs, but

* We will see why in chapter 6.

he could not do it and was frustrated. Bada suggested that he instead write the structure of methane, and both were stunned when Miller confidently and correctly wrote 'CH_4'. Despite the strokes, at least some of his vast chemical knowledge had survived.

Miller's career was over, but his experiment soon took on a new life. In 2007 his old lab had to be cleared out, and he had kept many of his samples – including those from the original experiments. By this time Miller was completely disabled, unable either to talk or to understand what was said to him. Nevertheless Bada visited him, not long before his death, and showed him one of the little boxes. Miller's eyes opened wide, as if in recognition.

Bada's team reanalysed the decades-old samples using modern techniques and determined that Miller had made more amino acids than anyone had realised – albeit in tiny amounts. In one experiment, Miller had tried to simulate the conditions in a water-rich volcanic eruption. The difference is subtle: one of the glass tubes is narrower, forcing more steam through the electrical current.[26] This variation created twenty-two amino acids, including some that are not found in proteins today.[27] The implication was that volcanic eruptions might have been factories for organic molecules on the young Earth.

What's more, in 1958 Miller had tried adding cyanamide: a white powder that resembles flour but is most definitely not good to eat. It's another simple molecule, made from five atoms: a carbon, two nitrogens and two hydrogens. This unassuming chemical may have helped drive the formation of crucial molecules like proteins. That's because cyanamide triggers 'dehydration' reactions: it causes other chemicals to shed a molecule of water. The formation of proteins from amino acids, and of nucleic acids like DNA, are both dehydration reactions. When Bada's team examined the cyanamide samples, they found a dozen amino acids, plus some that had linked up into pairs.[28]

Nobody knows why Miller did not publish this, but it was a prescient idea. In a barrage of papers published in 1977, Oró had shown that cyanamide drives the synthesis of many organic molecules, including simple proteins.[29] Furthermore, we will see in chapter 14 that cyanamide may be more crucial than Miller could ever have suspected.

This coda is pleasing, but the fact is that within a decade of Miller's seminal experiment, the whole thing was arguably irrelevant. The chemistry was being fought over, as the scepticism that greeted Ponnamperuma and Sagan's synthesis of ATP illustrates. Arguments had arisen over what the young Earth's atmosphere was like, and whether the step-by-step chemical syntheses being performed were really 'prebiotically plausible'.

But the bigger issue had emerged in the 1950s and 1960s. The discovery of the structure of DNA led to an explosion of understanding of the inner workings of living cells, and with it the realisation that life was far more complex than Oparin, Haldane and Miller could have been expected to comprehend in 1953. Making some amino acids and other organic molecules in a primordial soup was not the explanation of life's origin. At most, it was the first step along an unexpectedly long road.

PART TWO

Strange Objects

'Living creatures are strange objects, as men of all past ages must have been more or less confusedly aware.'

Chance and Necessity, Jacques Monod[1]

Chapter 4

The DNA Revolution

If you want to understand how life began, it helps to understand what life is and how it works. But in the early twentieth century, biologists arguably did not. In the 1920s, when Oparin first envisioned blobs of jelly floating in the primordial ocean, science's understanding of the internal workings of cells was extremely limited. It was possible to imagine a bacterial cell as being essentially a bag of enzymes and other chemicals: furiously busy, but perhaps not overly intricate. However, over the next few decades a series of major discoveries revealed that cells are far more finely tuned and multi-layered. At the centre of this story was the discovery of the structure and function of DNA. As biochemists learned more, it became apparent that DNA is the heart of an intricate molecular machine that exists in every living cell; a machine whose origins were extremely hard to fathom.

Today, DNA is famous. Mention those three letters and a cascade of images and ideas flow through people's minds: the beauty of the DNA molecule, notions of inheritance and parental legacy, genetic diseases, mutation caused by nuclear radiation, and perhaps 'designer babies' and grim stories like *Never Let Me Go* and *Gattaca*. The idea that our genes are made of DNA is part of our cultural lexicon.

But it came as a tremendous surprise to most biologists, who for the first half of the twentieth century were convinced that DNA was a sideshow: a molecule too simple to carry genetic information. Most biologists suspected instead that genes were carried by molecules called proteins, which are more complex than DNA.*

The first person to study what we now call DNA was Swiss biologist Friedrich Miescher, who in 1868 arrived at the University of Tübingen in what is now Germany.[1] He began trying to identify the chemicals in white blood cells, which he obtained by collecting human pus from surgical bandages. Miescher soon found something unexpected: an unknown milky-white substance resembling clumps of wool. It came from the nucleus of the cell, a blob-shaped region that appeared darker than the rest. So Miescher named his white substance 'nuclein', and published his findings in 1871.[2]

Miescher went on to study nuclein for a quarter of a century. He discovered that it contained carbon, hydrogen, nitrogen, oxygen and phosphorus. He also found that it was an acid, so its name was changed to 'nucleic acid', and that it was particularly common in the heads of sperm, suggesting it played some role in inheritance. But Miescher and his colleagues were unwilling to accept that a single substance might be responsible for inherited traits being passed on. Everyone thought this was too simple.

* We'll come back to proteins in a big way in chapter 7.

Reading his arguments with the benefit of hindsight is deeply frustrating: he came so close to figuring it out.

Inspired by Miescher's work, the German biochemist Albrecht Kossel spent much of his career breaking nucleic acid down into its component parts. Between 1885 and 1901 he and his students discovered that it contained five smaller chemicals, which we now call nucleobases.[3] These were adenine, cytosine, guanine, thymine and uracil.

The biochemist who arguably did most to unveil what nucleic acid really is, and to encourage other scientists to ignore it, was Phoebus Levene. Born in what is now Lithuania, his family emigrated to the US in the early 1890s to escape anti-Jewish pogroms. He was a short, thin man with the habit of wearing an extremely bedraggled hat. He spoke half a dozen languages, played the violin, and worked like a madman despite being rather frail.

Levene discovered that nucleic acid could be broken down into smaller molecules called nucleotides.[4] Each nucleotide was built from three components: a nucleobase, a sugar and a phosphate. Nucleic acid was a chain of these nucleotides strung together.

Furthermore, Levene found that there were two types of nucleic acid, which were distinguished by the type of sugar. Miescher's nucleic acid was not one substance, but two jumbled together. One used ribose sugar, so Levene called it ribonucleic acid or RNA. The other used deoxyribose, which is almost but not quite identical to ribose. This was deoxyribonucleic acid: DNA.

DNA and RNA also differed in another respect: they contained a different mix of nucleobases. Both contained adenine, cytosine and guanine – but where DNA contained thymine, RNA contained uracil.

Unfortunately, Levene made a suggestion that would ultimately prove incorrect. He had found that the four bases in DNA were always present in equal numbers, so he concluded that such a simple, repetitive molecule could not be the carrier of genetic information.

The first step towards the discovery that genes are DNA was taken by sheer accident by a British bacteriologist named Frederick Griffith. In the 1920s he studied a bacterium called *Streptococcus pneumoniae*, which had caused lethal secondary infections during the Spanish flu of 1918. Griffith found that the bacteria came in two forms: a strain with a smooth coat, which was infectious, and a rough-coated strain that was harmless. When he killed the smooth bacteria, they became harmless – as one might expect. But when he mixed dead smooth bacteria with living rough bacteria, the rough bacteria turned lethal and stayed that way for generations. A substance that conveyed lethality – what we would now call a gene – had escaped the dead smooth bacteria and entered the living rough bacteria. However, Griffith was shy, and slipped his astounding result out in an obscure journal in 1928.[5]

The person who picked up the finding and ran with it was Oswald Avery at Rockefeller University in New York. Small and bespectacled, Avery had hyperthyroidism, which caused his eyes to bulge out, until he had his thyroid removed in 1934. In the 1930s and 1940s, Avery's team repeated Griffith's experiment and searched for the chemical responsible. They meticulously destroyed or removed all the obvious suspects – and yet the leftover chemicals still transferred lethality. The only thing that made any difference was to add an enzyme known to break down DNA. Therefore, genes were made of DNA. 'Who could have guessed it?' Avery wrote to his brother. The team finally published in 1944.[6] However, despite Avery's thoroughness, his results were not widely accepted because the experiments were supposedly not rigorous enough.

The experiment that is often presented as settling the question was performed by Alfred Hershey and Martha Chase at the Carnegie Institution of Washington. They studied simple viruses that only contain DNA and protein.[7] Such viruses must infect bacteria in order to reproduce. Hershey and Chase found that during infection most of the viruses' DNA entered the bacteria, while the proteins largely did not – suggesting it was the DNA that was crucial. The experiment was less convincing than Avery's, as the pair had been less successful at eliminating contamination. Nevertheless, many were persuaded. When the results were published in September 1952, they triggered a race to figure out DNA's molecular structure and how it worked.

First on the scene was Maurice Wilkins, a British physicist who had worked on the Manhattan Project, only to abandon physics in disgust after Hiroshima and Nagasaki were devastated by nuclear bombs. In *Life's Greatest Secret*, Matthew Cobb describes him as 'quiet and reserved, with an odd habit of turning away from the person he was speaking to' and 'prone to suicidal thoughts'.[8] Wilkins had spent the latter half of the 1940s studying DNA at King's College London. To figure out its structure, he turned to X-ray crystallography. This entailed shooting X-rays at a sample of DNA, causing them to veer off in all directions. The resulting scatter pattern offered clues to the shape of the molecule, but interpreting it was fiendishly complicated.

In 1950, Wilkins's supervisor, John Randall, hired another X-ray crystallographer: Rosalind Franklin. Randall caused confusion immediately by writing a letter to Franklin in which he implied that she would be in sole charge of the DNA research – so when Wilkins returned from holiday and found her hard at work on his pet subject, he was aggrieved. This misunderstanding might have been resolved, but neither was good at communicating. Where Wilkins was quiet, Franklin was forceful and brusque.

Making matters worse, Franklin also encountered sexism. Accounts differ on just how bad things were. Franklin's friend Anne Sayre claimed in a biography that few female scientists were hired and they had to eat lunch in a different room from the men.[9] However, these details have been disputed and other accounts suggest Franklin's department was more accommodating of female scientists than the rest of King's.[10] But things were clearly far from ideal: women were excluded from after-lunch coffee in the smoking room, so missed key discussions.[11]

The third member of this dysfunctional group was James Watson, who had obtained his PhD at the age of twenty-two. Watson had been persuaded to get into DNA research by the Hershey–Chase experiment, and had seen Wilkins give a talk in which he presented some X-ray crystallography images of DNA. He wound up sharing an office with Francis Crick, another physicist-turned-biologist, at the Cavendish Laboratory at the University of Cambridge. Crick was in his thirties but was still trying to get his PhD, a previous project having been interrupted by the war – in particular by a bomb that fell through the roof of his laboratory and destroyed all his equipment.[12] Watson and Crick immediately found each other's company electric. They determined to figure out the structure of DNA first, preferably without doing any actual experiments.

The competitors were assembled and the race was on.

On 21 November 1951 Franklin presented some new images of DNA, which suggested that the molecule had a spiral structure and that, if it was made up of multiple strands, they ran in opposite directions. Watson was in the audience, but he didn't take any notes and – by his own account – instead mused on Franklin's looks, so he could only give Crick an incomplete account. As a result the pair's first model was completely wrong. It featured three chains of sugars and phosphates, intertwined in a helical shape, with the bases sticking out like spikes. They embarrassed themselves by inviting Franklin and Wilkins over to show it off, only for her to instantly explain why it did not match the X-ray data.

Things dragged on through 1952. Franklin took ever more X-ray images and gradually convinced herself that Watson and Crick were right about one thing: DNA was some sort of helix. But she was unsure how many strands there were. Meanwhile, Crick and Watson spent much of the year on other things. They were only prodded back into action in January 1953 when an American group claimed to have the solution – which the pair were relieved to discover was wrong.

Recommitting, Watson visited King's. After another row with Franklin, he visited Wilkins in his office – and was shown one of Franklin's images, which was particularly clear and confirmed to him that DNA was a helix. This is perhaps the most controversial incident in the story. Arguably Wilkins should not have shown him the photo, at least not without asking Franklin. Then Crick was shown a written report that described Franklin's 1951 results in more detail and he finally understood that DNA was made up of two chains, running in opposite directions. Finally, in

February Watson figured out how the four bases fitted together inside the helix: adenine could only pair with thymine, while cytosine could only pair with guanine. This was crucial, because it offered a way for DNA to copy itself and keep the sequence of bases intact. The pair then worked furiously to turn this insight into a precise model. They scooped Wilkins, who was nowhere near the solution, and Franklin – who was agonisingly close, and whose results had been crucial.

The structure turned out to be disarmingly simple. There are two strands, which twist around each other like the coils of a rope, or two spiral staircases that run in opposite directions. Each strand is made of alternating sugars and phosphates. The two strands are linked by bridges, each of which consists of two bases: either adenine and thymine, or cytosine and guanine. It can be helpful to imagine the molecule as a ladder, with the bases as the rungs, except that in reality the ladder is twisted into a spiral.

Watson and Crick's paper concluded with a cocky, teasing sentence: 'It has not escaped our notice that the specific pairing we have postulated immediately suggests a possible copying mechanism for the genetic material.' The idea, which they thought so obvious they didn't bother to spell it out, was that the DNA molecule could split into its two individual strands. Each of these could then hoover up loose nucleotides floating nearby. The newly formed strands would necessarily always have the correct sequence of corresponding bases, because any incorrect bases would simply not fit.

Watson and Crick's structure of DNA was published in *Nature* in April.[13] Their short paper was accompanied by articles by Wilkins and Franklin, describing the X-ray data.[14] Fraught and perhaps disreputable as the process of figuring out the structure had been, the result was one of the greatest scientific discoveries in history. The molecule that underpinned heredity had been unmasked, revealing how organisms pass on their traits to their offspring. Any hypothesis about life's origin would now have to explain the origin of DNA.

The media largely ignored it. Whereas Miller's study earlier that year had received glowing coverage, the work of Watson, Crick, Franklin and Wilkins passed almost unnoticed. But it did slowly sink in, and Crick, Watson and Wilkins shared the 1962 Nobel Prize in Physiology or Medicine.[15] Franklin had died of ovarian cancer four years before, aged thirty-seven.

With hindsight, some of the ideas underpinning Watson and Crick's discovery were in the air. Both had been influenced by the physicist Erwin Schrödinger,* who suggested in his 1944 book *What Is Life?* that genetic information must be stored in an 'aperiodic crystal': that is, a crystal with a molecular structure that had some sort of variation, rather than simply repeating the same set of atoms over and over.[16] Similarly, in 1927, Soviet scientist Nikolai Koltsov suggested that the genetic molecule might be 'two mirror strands that would replicate'.[17] Nobody followed this up and, predictably, Koltsov was denounced by Lysenko and in 1940 was fatally poisoned by the Soviet secret police. However, Watson and Crick probably never saw his idea.[18]

*

* As in Schrödinger's Cat, a thought experiment about a cat in a box. The experiment mostly demonstrates that Schrödinger did not know much about cats, because if he had he would have understood that it's almost impossible to put a cat into a box.

With the structure of DNA established, the next challenge for biochemists was to figure out what DNA actually did. What was the message contained in the sequence of bases? Once again, Crick supplied the answer: it was the instructions for how to build proteins, which are some of the most ubiquitous and important molecules in living organisms. By spelling out which proteins would be made, DNA could control the inner workings of cells. We will take a closer look at proteins in chapter 7, but for now the key point is that proteins are chains of smaller molecules called amino acids. Life only uses twenty amino acids, but they can be strung together into chains that are hundreds of amino acids long, like letters in a long word such as 'pneumonoultramicroscopicsilicovolcanoconiosis'. Somehow, the four bases in DNA were encoding the correct sequence of amino acids.

The next task was to crack this 'genetic code'.* The problem was twofold: decrypting the message in the DNA, and figuring out the mechanism by which it was used to make proteins. The scientists pursuing this were not necessarily interested in life's origin, but origins researchers paid close attention: they knew they would need to explain how the DNA–protein system got started.

How might DNA encode the sequence of amino acids needed to make a protein? It is a bit like translating between languages

* Arguably, it isn't a code at all, but a cipher. Codes work at the level of meaning, so replacing the word 'fox' with a fox emoji would constitute a code. In contrast, ciphers replace or jumble up the individual letters: if you were to replace all the letters in a word with the next highest letter in the alphabet, turning 'cat' into 'dbu', that would be a cipher. Unfortunately, pointing out this distinction will rarely impress people at parties.

with different alphabets. The DNA alphabet consisted of the bases that were strung along the molecule. There were only four bases, which could be thought of as letters: A for adenine, C for cytosine, T for thymine and G for guanine. However, the protein alphabet had twenty amino acid 'letters'. The challenge was to find a way to represent each of the twenty letters of the protein alphabet, using only the four letters of the DNA alphabet.

The simplest imaginable system was that each base coded for one amino acid, but this could never work as only four of the twenty amino acids could be represented. Instead, each amino acid would have to be represented by a short string of bases. But how many? Pairs of bases such as AC or GT would also not do, because such a system permits just sixteen combinations (four multiplied by four), which is still not enough to cover all twenty amino acids. So there would have to be three bases for each amino acid. Three bases gave sixty-four combinations, far more than seemed necessary, but there was no other way. The idea that the code was based on triplets was debated for years, but proved correct.

One of the first to have a crack at how this worked was the Russian-born physicist George Gamow, who is better known for being one of the first to propose that the universe began with the 'Big Bang'. In 1954 Gamow put out a short paper establishing the idea that the DNA bases worked as a cipher.[19] He described proteins as 'long "words" based on a 20-letter alphabet' and asked how 'four-digital numbers can be translated into such "words"'.

Gamow went on to form an eccentric group called the RNA Tie Club, named for DNA's sister molecule, devoted to cracking the genetic code. The club counted Watson and Crick among its twenty members, each of whom was assigned an amino acid and given a woollen tie emblazoned with an illustration of an RNA molecule. They spent a lot of time on semi-drunken brainstorms.

Unfortunately, Gamow's chemical ideas were wrong. He assumed proteins were assembled on the DNA itself, and imagined a lock-and-key mechanism in which the amino acids fitted into the gaps between the nucleotides. However, it was already clear that proteins are not made on DNA. Quite where they *were* made started to emerge in 1955, when Romanian biologist George Emil Palade spotted tiny granules inside cells.[20] Soon dubbed 'ribosomes', the granules turned out to be enormously complicated molecules containing both RNA and protein. Gradually, it emerged that the ribosomes were the place where amino acids were joined into proteins. Somehow the instructions were being taken from the DNA, which was stored elsewhere, to the ribosomes.

On 19 September 1957, Crick gave a lecture at University College London in the UK. He published it as 'On protein synthesis' the following year.[21] That title, echoing Darwin's thesis on evolution, would seem self-important if Crick's ideas weren't brilliant.[22]

Following Gamow's lead, Crick didn't worry about the still-mysterious chemistry of how proteins might be made, but instead described it as a flow of information. There is information in DNA, stored in the sequence of bases, and this information is somehow 'translated' into a different language, that of amino acids. Crick dubbed this idea, that information can be carried from DNA into protein but not in the other direction, the 'central dogma'. He came to regret this name, because it implied that the idea should never be challenged, and later admitted he had not fully understood what 'dogma' meant.

Furthermore, Crick correctly proposed the sequence of events by which proteins were made. First, a gene is copied from DNA onto an RNA, which carries the information to the ribosomes. The ribosome then makes a protein, using the information on the RNA as a guide. Crick suggested that the amino acids were physically assembled on the RNA 'template'. But how would each amino acid know where to go? His solution was another kind of RNA, which he called an 'adaptor'. Each amino acid would be carried by a different adaptor RNA, and the adaptors would attach themselves to the template RNA by base-pairing, just like in DNA. This would arrange the amino acids in a row, in the correct sequence.

Crick's foresight here was staggering. Even as he gave his lecture, researchers at Harvard University were isolating the adaptors, which came to be known as transfer RNA.[23] Each transfer RNA is a single strand, folded into a shape that resembles a clover leaf. Just as Crick had suggested, a different transfer RNA carries each amino acid.

Four years after Crick's lecture, another form of RNA was discovered. This was the molecule that carried a copy of a gene and acted as a template, on which the ribosome could assemble a string of transfer RNAs and thus build a protein. It was soon dubbed 'messenger RNA'. Two groups, one of them including Watson, described it simultaneously.[24]

All the pieces were now in place. When a protein needed to be made, the relevant gene was copied onto a messenger RNA. This new molecule puttered over to a ribosome, where it was used as a template to make a protein. Amino acids were assembled in a line by transfer RNAs, which base-paired to the messenger RNA. Finally, the ribosome stitched the amino acids together. The physical mechanism of how proteins are formed was becoming clear.

However, in eight years no progress had been made on decrypting the sequence of bases in DNA. That meant nobody knew which sequence of bases coded for which amino acid, so nobody could predict which bit of DNA would produce which protein. The language of DNA still could not be read. Crick and other prominent thinkers had tried hard, to no avail, only for two outsiders to swoop in and crack the problem.

The experiment that read the first word of the genetic code was devised by Marshall Nirenberg at the National Institutes of Health in Bethesda, Maryland. He was not part of the inner circle of genetic code researchers and had even been refused admission to a conference on the subject. On 27 May 1961, his colleague Heinrich Matthaei ran the experiment. Matthaei created an RNA that only had one type of base, as this was the easiest kind to make. He chose uracil (U), so the RNAs had sequences like UUUU or UUUUUUUUUU. When he fed these into a ribosome in a test tube, the result was a protein made entirely of one amino acid: phenylalanine. Clearly, a sequence of Us coded for phenylalanine. Once it became clear that Gamow had been right and the code uses three bases for each amino acid, it was possible to be specific: the code for phenylalanine is UUU.

When Nirenberg and Matthaei announced their findings later that year, it triggered a frantic race to decrypt the other sixty-three triplets, which were dubbed 'codons' the following year.[25] By 1967 all sixty-four codons had been decrypted.[26]

The unravelling of the genetic code was the culmination of two decades of profound discoveries that would revolutionise the life sciences. From the structure of DNA, through the existence of ribosomes, messenger RNA and transfer RNA, and finally the genetic code itself, biochemists had uncovered processes that lie at the heart of all life. In the coming decades, scientists would read the entire DNA sequences or 'genomes' of hundreds of species, including humans. They would discover genes that underlie diseases, and create genetically modified organisms, including humans with edited DNA. The consequences of the DNA revolution are still playing out.

But one consequence was immediately apparent, or ought to have been. For everyone trying to figure out how life began, this cascade of discoveries was a cluster bomb. Cells were not just blobs of amorphous, jelly-like protoplasm: they were furiously intricate machines with assembly lines manufacturing proteins based on molecular blueprints.

And there was much more besides. As we'll see in chapter 10, cells perform sequences of chemical reactions to obtain energy from chemicals in their environment, a process called 'metabolism'. Even the simplest of these mechanisms have multiple steps. Cells also have a kind of internal scaffolding. This was first suggested in 1903 by Nikolai Koltsov,* and by the 1990s it was clear that even bacteria have it. Even the outer walls of cells are not simple barriers. They are dotted with large molecules that draw desirable chemicals into the cell and shoo out waste material.

In short, even the simplest living cell is incomprehensibly complex, with hundreds of genes interacting to build and control a host of moving parts. There was no way such an intricate

* Yes, the same Nikolai Koltsov who anticipated base-pairing a quarter of a century before Watson and Crick thought of it.

system could have arisen in one fell swoop. The odds were beyond astronomical. Worse, many of the parts were irreplaceable: for instance, a cell without ribosomes would die.

A few origins researchers understood this quickly. One of them was John Desmond 'J. D.' Bernal. Like Haldane he was a communist and brilliantly creative, with interests ranging from the study of crystals to the origins of life – he also helped plan the Normandy landings. Later he employed Rosalind Franklin after she left King's College London.

Bernal could not attend the 1963 conference where Haldane and Oparin met, but he sent an essay in which he offered a stern warning.

> We have to face much greater difficulties than we had thought earlier we had to face, because it is becoming more and more evident, now that something is known about the mechanism of nucleic acid reproduction and the complicated stages of the production of enzyme proteins, that many more stages must have intervened than we had previously thought necessary. The very perfection of these mechanisms, however, makes it more difficult to see how they could have come into existence spontaneously together with their functions.[27]

What was needed was some lateral thinking. And a Scottish chemist named Graham Cairns-Smith was on hand to provide it.

Chapter 5

Crystal Clear

The late Graham Cairns-Smith (1931–2016) is perhaps the most unusual person featured in this story. His ideas have a peculiar status in the origin-of-life community: something of a curiosity, but not to be dismissed entirely. His most significant work was done in the 1960s and in many ways it was decades ahead of its time. Reading his papers now, there is a sense of something almost there, a truth glimpsed out of the corner of the eye.

I never met Cairns-Smith, but I feel an odd kinship with him because we crossed paths at the end of his life. In 2016 I was editing a website called BBC Earth and commissioned freelance writer Martha Henriques to write a story about Cairns-Smith's research. It was the fiftieth anniversary of his first major paper on the origin of life, and this seemed an ideal time to reassess his ideas. We also knew that Cairns-Smith was eighty-four, so this might be the last chance to speak with him.

In June, Martha travelled to Glasgow to interview Graham and his wife Dorothy at their home. Graham was ailing, stricken by a Parkinson's-like disease that was affecting his mobility, but he was able to speak lucidly and at length. We ran Martha's story on 24 August and readers were fascinated.[1] But a few days later Martha emailed to say that Dorothy had been in touch. Graham had died on 26 August. His interview with Martha had been his last. Our story had come out just in time for him to see it, and Dorothy said it had pleased him. We added a line announcing Graham's death.

Editing this story was a peculiarly emotional experience. I felt a strange closeness to Graham, despite never having met him. Martha and I were looking at his entire life, albeit in outline, and I felt a weight of responsibility to do it justice.

Alexander Graham Cairns-Smith was born on 24 November 1931 in Kilmarnock in the south-west of Scotland. It's a town best known for its football team – so much so that when you search for it on Google, the club's official website ranks above the Wikipedia entry for the town. He went to a series of schools, ultimately ending up at Fettes College in Edinburgh. This boarding school is occasionally called 'the Eton of the north' by journalists, mostly as a term of abuse. Still, Fettes nurtured Cairns-Smith's twin loves: science and art.

Being a boarding school, Fettes organised extracurricular activities on weekends, including art classes on Saturdays. But these were no ordinary art classes, and Cairns-Smith learned from three top-quality mentors. The first was Robert Heriot Westwater, whose somewhat Impressionist style was influenced by Degas. The second was William 'Willie' Wilson, whose parents were evidently fond of the letter W and who variously made stained-glass windows, prints and watercolours. The third was Denis Peploe, a landscape artist and sculptor.

With their guidance, Cairns-Smith got onto the radar of the Royal Scottish Academy in the 1940s, while still in his teens. Becoming a professional artist seemed a real possibility.

Most of his works were oil paintings, often landscapes – although he went through at least one abstract phase. Today, a digitised version of one of his pieces is held online in the Argyll Collection, which represents artists from that region. Cairns-Smith's piece bears the firmly minimalist title, *A landscape with fields*.[2] It is an aerial image of the sort you might see looking out of an aeroplane, rendered in thick brushstrokes. Ploughed fields are sludgy oranges and reds, contrasted against near-black tree lines. Any art historian would instantly place it in the mid-twentieth century, and a good one could probably take a stab at the decade.

Cairns-Smith was also no slouch at poetry. When he was at Fettes he contributed humorous poems to *The Wart* (nowadays *The Hive*), a scurrilous magazine run by the pupils.* Couple that with his talent for wine tasting, for which he won a national prize, and it becomes clear that he was something of a Renaissance man.

He had loved science since childhood. His maternal grandfather had been an amateur astronomer, and his mother passed this interest on to him. By his late teens he had to make a choice between science and art, and he did so with admirable practicality. There was no way to be an amateur scientist in the modern

* As a veteran of *Clareification*, the in-house student magazine of Clare College in Cambridge, I can attest that the contributors to such publications have never seen an abyss of bad-taste humour that they didn't immediately plunge into.

era: research requires so much equipment and education. But he could be a professional scientist and amateur artist. Science also offered a more stable life. It is hard to imagine someone making the same decision today: if the artist's lifestyle is erratic, the young academic's is peripatetic and cash-strapped. But this was a different time, with student grants and the prospect of early tenure. Cairns-Smith claimed not to have regretted the decision, but at least one of his tutors did. The painter William Crosbie, who taught him a few classes, muttered that it was 'a pity you chose science'.

Cairns-Smith graduated in chemistry from the University of Edinburgh in 1954. He immediately began a PhD, which he completed in 1957, before becoming an assistant lecturer at Glasgow University.

By 1961 he had become a warden of one of the schools of residence, where he met Dorothy Anne 'Dodo' Findlay. 'I saw him across a crowded room,' Dodo recalls. 'I thought his shirt looked nice. I asked him if he'd like to come back for supper and he said no.' Later, Dodo saw Cairns-Smith walking with another man and this, along with his good taste in shirts, led her to conclude that he was homosexual.

However, the pair were brought together again at a party thrown by mutual friends. Dodo wanted to be set up with someone and a friend suggested Cairns-Smith. 'I said it's no good, he's gay,' she says. They finally bonded over their common interests in music and wine, and Dodo secured his attentions by placing an expensive case of Château Latour under her bed. They married on 17 August 1962, had three children, and were together until his death.

*

While Cairns-Smith had been studying, biology and organic chemistry were being revolutionised. As we saw in chapter 4, Watson and Crick published their double-helix structure of DNA in 1953: the year before he graduated. Five years later, Crick set out his 'central dogma': DNA makes RNA, which carries the instructions for making proteins. This set off the race to decrypt the genetic code: the sequence of 'letters' carried by DNA and RNA. The deepest secrets of life were being uncovered.

The year 1953 also saw the publication of the Miller–Urey experiment, which presented hard evidence that the molecules of life could form naturally from simple chemicals, as we saw in chapter 3. This offered to explain how life began.

To Cairns-Smith the experiment was 'deeply interesting' and 'beautiful'. But it was not even close to an explanation for life.

Miller had made a handful of amino acids, the building blocks of proteins. But living things use twenty amino acids, allowing them to make a staggering diversity of proteins. They also use DNA and RNA, not to mention sugars, lipids and a host of other chemicals. Perhaps the first life did not use the full gamut of chemicals that modern life does, but, even so, Miller's amino acids were surely only a fraction of what was needed for life to form. Worse, he had made them in minuscule quantities, and they were jumbled up with dozens of other chemicals that were of little or no use to life.

Cairns-Smith's studies in organic chemistry had also taught him another harsh fact: the molecules of life are delicate. If you put DNA in water at room temperature and leave it to its own devices, it will swiftly fall to pieces. It survives in our cells thanks to helper molecules that shield it and repair it when it breaks. Even if Miller's experiment could be refined to produce a wide range of amino acids and whatnot, they would never assemble into proteins and DNA – and, if by some miracle they did, these large molecules would soon fall apart again.

In the early 1960s, as Martin Luther King led the civil rights movement to Washington DC and The Beatles came tumbling out of the dive bars of Hamburg to conquer the world, Cairns-Smith spent four years reading everything he could find that was relevant to the origin of life, from geology to biology and chemistry. The university let him get on with it, and if they cared that he wasn't producing any new findings, they showed no sign. Cairns-Smith taught and supervised students, and spent time with his new family, but otherwise he was quiet.

By 1966, Cairns-Smith had found his central idea. He would pursue it for the next twenty years. His insight was to step back from the fiddly molecular details of DNA, proteins and the rest, and ask a deeper question: what are the fundamental properties of life? In other words, what does something need to do in order to be alive?

In the scientific world, if not in the classrooms of America, there is and was consensus: life evolves. A living organism's main goal is to reproduce, passing on its genes to its offspring – preferably in such a way that its offspring in turn will have a good chance of passing on their genes. But with each generation, the genes change. Mutations pop up, largely as the result of mistakes as the genes are passed down. If a mutation is beneficial, in the sense that it makes its host organism more likely to reproduce, it thrives and becomes more common in the population. If it is harmful, it becomes less common and may vanish entirely.

At its root, then, life was an exercise in the imperfect copying of information, a game of Chinese whispers played with molecules instead of spoken words. If you stripped away all the frippery, all the peacock tails and beating hearts and chattering tongues, each living organism was nothing but a set of instructions with mistakes in it. The modern version of this exercise in sloppy transcription involves DNA and the other molecules we associate with life. But perhaps in the beginning it used something simpler and less fragile, something that was present on the young planet Earth in huge quantities.

The material Cairns-Smith finally alighted on was utterly counterintuitive: clay. Those of us who were hopeless at pottery in school may not have seen anything terribly wonderful in clay, but if you look at it under a microscope this sticky mess reveals itself as a kind of wonderland. It is made of tiny crystals in a dizzying array of shapes.

These crystals can be thought of as somewhat alive. They grow if you place them in water laced with the right chemicals. They also split apart, so a 'mother' crystal gives rise to smaller 'daughter' crystals. Cairns-Smith reasoned that they might even be able to evolve. After all, a mother crystal would have its own peculiarities: a particular chemical makeup, perhaps, or distinctive cracks. These could be passed down to its 'offspring', just as we pass on our genes to our children. Similarly, when a crystal split into two, new changes might occur; perhaps one of the daughter crystals would pick up a new chemical from the surroundings. This would be analogous to a new mutation.

None of this means that the crystals are actually alive: they simply have some of the properties of life. But these non-living crystals could pass on advantageous and disadvantageous traits to their offspring, just like living organisms. Perhaps some properties would make it easier for a crystal to split in two, allowing it and its descendants to reproduce faster – causing it to become more common. Or perhaps a certain chemical would weaken a crystal so much that it disintegrated under the slightest pressure.

In short, an entire evolutionary history could have played out in wet clays on the early Earth. Later, the clay crystals could have started to make use of 'biological' molecules such as DNA and proteins, perhaps using them to store additional information or to carry out other functions. Eventually, these helper molecules took over entirely and left the clay behind – as if an archway, once erected, chose to dismantle its own scaffolding.

Cairns-Smith set out his ideas in a paper in the *Journal of Theoretical Biology*, published in January 1966.[3] The paper feels like an opening salvo in a hugely creative year, which also gave us the Beach Boys' *Pet Sounds*, Bob Dylan's *Blonde on Blonde* and The Beatles' *Revolver*. Its ideas are expressed with startling clarity. Cairns-Smith used vivid analogies like detonators and starting guns – essential in a paper that fused the disparate fields of evolutionary biology, chemistry and mineralogy.

It is a profoundly elegant hypothesis, and it is no surprise that it found many fans. The evolutionary biologist and writer Richard Dawkins devoted much of a chapter to it in *The Blind Watchmaker*.[4] After mentioning the then-predominant idea of a primeval soup, Dawkins – ever the contrarian – announced that he wished to 'fly a kite for a somewhat less-fashionable theory', the 'audacity' of which he admired.

Yet Cairns-Smith's ideas have never found widespread scientific acceptance. They don't turn up in textbooks of biology, except perhaps as a historical curiosity, and have barely been followed up. 'People dismissed him at the beginning as being completely mad as a brush,' says Dodo. Miller was particularly unimpressed.

'Stanley was always very disparaging about Graham. He always said he was a loser.'

The problem is simple: the hypothesis is fiendishly hard to test. One would have to identify individual crystals within a clay and note their properties, then follow their fortunes as they grew and divided, and the fortunes of their offspring. Simply looking under a high-powered microscope would not do the trick: new experimental techniques would be needed. Worse, those techniques might only be useful for studying clay. Similar batteries of techniques have been developed to tackle other problems, such as sequencing the human genome – but that project promised huge benefits to society, such as curing genetic diseases. Testing Cairns-Smith's hypothesis would be an exercise in pure curiosity.

As a result, we don't know if clay crystals 'evolve' in the way Cairns-Smith described. He never managed to demonstrate it, and for twenty years after his initial paper he was confined to fleshing out and restating his ideas. Despite trying repeatedly, he never obtained the funding necessary to begin experimental trials.

This took its toll. In the 1970s, Cairns-Smith suffered a bout of depression that lasted a year. 'He suddenly thought, "Maybe everyone's right and I am wrong,"' recalls Dodo. 'He thought, "I've wasted all these years."' In 1972 he also gave up painting, because there was no longer room in the house for him to do it. Less than a decade before, he had been putting on one-man shows at the prestigious McLellan Galleries in Glasgow: Dodo recalls that at his first show, around 1965, he displayed thirty-six paintings and sold thirty-two. But now it was taking up too much of his mental energies, and he decided to focus on writing.

The same skill as a writer that made his 1966 paper so compelling would serve him well in keeping his ideas alive. His first book, *The Life Puzzle*, came out in 1971. He followed it in 1982 with *Genetic Takeover and the Mineral Origins of Life*. Neither swayed his critics. A reviewer of the second book complained that it contained too many expressions like 'might have been' and 'would have been possible', and little that was 'concrete'.[5]

However, by the early 1980s it seemed there might finally be some experimental evidence for his ideas. In 1980, a chemist named Armin Weiss at the University of Munich in Germany spoke at a conference, where he described experiments with clays that seemingly showed they could indeed replicate. He published his results the following year.[6]

The experiments used a kind of clay called montmorillonite,* which can form thin layers like a stack of pancakes. Weiss claimed that when a new layer formed, sandwiched between two existing layers, information was copied to it. The experiments did not image the molecular structures of these clay layers, but they did show that the distribution of charged particles held fairly steady from one layer to its daughters. After twenty generations the distribution became unrecognisable. Nevertheless, Weiss claimed the layers had replicated successfully for a handful of generations. He called the clays 'an excellent replicating system model' that behaved like 'the most primitive forms of protolife'.

However, nobody could reproduce Weiss's results. They asked

* It isn't particularly important here, but readers would do well to remember the name montmorillonite. Think of it as a gun hanging on a wall, ready to go off at a crucial point later in the story.

him for details but could not get satisfactory answers, so became suspicious that the results were fraudulent. '[Leslie] Orgel and Miller wanted to expose him,' recalls Dodo. Ultimately, five origins researchers co-authored a response.[7] Cairns-Smith, to his credit, was among them. He removed the suggestion of fraud, but the reply was still (by academic standards) blistering. 'The information given in the Weiss 1981 paper is altogether inadequate for others to repeat the key experiments,' the researchers wrote. Weiss's paper was a review, compiling existing results. The key results were seemingly held in three theses, produced by students. None had been published in a scientific journal. 'We have been unable to trace the first and third of these theses,' the researchers complained, 'while the second contains nothing which we consider of special relevance to clay replication.'

Weiss never did publish anything further on the subject.[8] Yet the incident does not seem to have unduly affected his career. When he died in 2010, his scientific obituary made no mention of anything untoward.[9]

Cairns-Smith's biggest popular success came in 1985 with *Seven Clues to the Origin of Life*.[10] Whereas his previous books were technical, this was aimed at a general audience and written as a detective story. Every chapter was bookended with quotations from Sherlock Holmes. Cairns-Smith rammed home just how difficult it is to imagine a living cell evolving from scratch. Everything is too perfectly put together. Proteins, DNA and RNA are all interdependent, so if one is removed the whole thing collapses. Instead, Cairns-Smith asked readers to imagine 'an arch of stones'. 'How can you build any kind of arch gradually?' he asked. 'The answer is with a supporting scaffolding' – by which he meant clay.

Reviewers praised the book's vivid style.[11] However, scientists remained unswayed, complaining that Cairns-Smith could not explain the transition from clay to biochemicals like protein.[12] Ultimately, *Seven Clues to the Origin of Life* did not get Cairns-Smith the resources to test his hypothesis, nor did it lead to a flurry of experiments by anyone else.

Instead, in 1986, events took a cruel turn. The Cairns-Smiths' son, Adam, who was close to completing his degree at the University of Edinburgh, took his own life. Grief-stricken, and with his experiences of depression playing on his mind, Cairns-Smith became obsessed with the nature of consciousness. 'It was really about, where do these feelings of despair and lust and so on come from?' Dodo says. 'There doesn't seem to be any place physically where they could be, but there they are.'

During the 1990s Graham produced two books, *Evolving the Mind* and *Secrets of the Mind*, in which he attempted to explain the so-called hard problem of how consciousness is produced by the brain.[13] There is a distressing tendency for highly creative scientists, late in their careers, to assail the consciousness problem, and it rarely ends well. Cairns-Smith's proposed explanation for the conscious mind relied on the peculiarities of quantum physics, and while a few other thinkers have followed a similar tack, so far the idea hasn't been blessed with experimental evidence. The consensus is that we shouldn't try to explain one mysterious thing by invoking a second mysterious thing. Reviewers dismissed Cairns-Smith's arguments as flimsy.[14]

That was about it for Cairns-Smith's scientific career. He and Dodo settled into a long, peaceful retirement. Before he died, he insisted that the order of service at his funeral display his date of birth, followed by 'best before 26 August 2016'.

*

What are we to make of Cairns-Smith's ideas on the origin of life? From a scientific point of view, they are hard to classify. They have never gained widespread acceptance, but they have also never been definitively dismissed. Certainly, there is no sense that Cairns-Smith was a crank or anything close to it. In many ways he was simply decades ahead of his time, with an inspired idea that could not be tested because there wasn't the technical capability to do so.

The first thing to say is that, since the ideas have never truly been tested, they are only an intriguing hypothesis. No matter how compelling the logic seems, without experiments we should not put much stock in it. The entire point of experimental science is that we are good at misleading ourselves and failing to notice the flawed assumptions behind an idea. This is particularly true when an idea is as intricate as Cairns-Smith's. There are so many steps that have to be true, from the properties of clay crystals to the supposed genetic takeover, that there is almost certainly a nasty trap somewhere. Only by testing the steps rigorously could we discover whether Cairns-Smith was on the right track.

However, we do not have to believe he was right to see that his work was of immense value. Cairns-Smith was working at a time when genes could not be read, let alone entire genomes. Many crucial fossil discoveries, especially from life's earliest stages, were decades into the future. When you consider that, it becomes clear that his hypothesis was a major intellectual achievement, for two reasons.

First, he moved beyond chemistry, which at the time dominated everyone's thinking about the origins of life. Cairns-Smith showed that it was not enough to make biological chemicals: one had to think about how they worked, and how those processes might happen in the context of the Earth, not just in a glass apparatus in a lab. He forced origin-of-life researchers to think seriously about rocks, sediments and water. Nowadays, any hypothesis on life's beginnings that did not do so would be laughed out of court, but at the time this was a leap forward.

His second achievement was to consider the fundamental nature of life, rather than the physical form it happens to have adopted on Earth. Without speculating about silicon-based life or the like, there is nothing in modern science to tell us that life has to harness exactly the molecules it does on Earth. Science has created alternatives to DNA, and there are plenty of amino acids that could be used to make proteins but which earthly life unaccountably ignores. Cairns-Smith cut through this detail to see that life is about information that is inherited imperfectly. That is something any good theory on the origins of life must handle.

It is perhaps crucial that Cairns-Smith was an artist as well as a scientist. His scientific ideas have a fantastic inventiveness. It must have taken a flash of true inspiration to see the similarities between genes carried by DNA and a clay crystal dividing in two. That ability to see past an object's surface detail to its underlying properties is one of the marks of true creativity, in art and science. The resulting idea, while it may well not pan out, echoes through the rest of this story. In later chapters we will see biochemists wrestling to produce a replicating molecule and geologists seeking the perfect cradle for life. Both groups were following in Cairns-Smith's footsteps. If some ideas are so bad they are 'not even wrong', as the physicist Wolfgang Pauli once complained,* other ideas may be valuable precisely because they are wrong.[15]

Cairns-Smith's greatest legacy may not be his actual ideas, but the effect they had on others. Think of him as the Velvet Underground of the origin of life: not many people read his ideas but everyone who did went straight to the lab.

* This is such a good line, the physicist Peter Woit used it for the title of a 2006 book attacking string theory, and subsequently for his blog. These days it is so much associated with Woit that it is easy to forget it was Pauli who said it.

Chapter 6

The Schism

Cairns-Smith's clay hypothesis was an attempt to move beyond the Oparin–Haldane primordial soup, to find a story that could explain the newly revealed intricacy of life's inner workings. But it never found widespread acceptance. Instead, over the 1960s and 1970s, the origin-of-life field splintered. Where once there was a rough consensus that Oparin, Haldane and Miller had been on the right track, there was now discord; dozens of competing ideas, their supporters heading off down different paths through the wilderness. Many questions that had previously been glossed over were now viewed as urgent. Ultimately, these would lead to the formation of several competing hypotheses for life's origin.

Not that most people would have noticed at the time. Outwardly, origins researchers stuck to the script, regularly telling journalists the solution was in sight and the broad strokes of the picture were correct. Carl Sagan's 1980 TV series *Cosmos* happily restated the primordial soup hypothesis without even a hint of the problems.[1]

It is perhaps no wonder that some researchers went into denial, maintaining a confident public front while privately flailing. The American biochemist Robert Shapiro knew better, however.[2] 'The effect of such publicity is simply to reinforce the credibility gap that has existed between the origin-of-life field and much of the rest of science,' he wrote. Pretending you know the answer when you clearly don't just makes you look silly.

There were two festering problems with the primordial soup hypothesis as outlined by Oparin and Haldane, and with Miller's experimental support for it.

The first issue concerned ultraviolet radiation. The Sun blasts out rather a lot of this, but these days the ozone layer keeps much of it out – which is good news for us, because too much ultraviolet is harmful. But ozone is a molecule made of three oxygen atoms, so only forms if there is oxygen in the air. Through the 1960s and 1970s studies of ancient rocks confirmed that when Earth was young there was no oxygen in the air, and therefore no ozone.[3]

Oparin had made creative use of this, suggesting that the ultraviolet provided energy for the chemical reactions that created the first organic molecules. Some of the post-Miller research suggested he was right – and we'll see in chapter 14 that evidence for this has only accumulated. But ultraviolet also breaks biological molecules: that is why it is harmful. So while ultraviolet may have helped biological molecules form, those molecules would have found themselves in a race to achieve something before they were obliterated.

Some researchers now believe life began deep in the sea, insulated from harmful ultraviolet. However, this raises the question of how the key molecules formed, without ultraviolet to drive the chemical reactions.

Meanwhile, others think life must have begun in the shallows or on land, and that the benefits of ultraviolet outweighed the dangers. Sagan was sympathetic to this argument. At the 1963 Wakulla conference, he pointed out that living organisms are surprisingly well-adapted to ultraviolet radiation, even though the protective ozone layer has existed for hundreds of millions of years.[4] This suggested to him that life arose in 'an ultraviolet-rich environment' and quickly evolved the necessary protections.

This argument about ultraviolet underscored a broader problem. It was not enough to show that chemicals like amino acids could have formed on Earth: it was also necessary to show how they might survive. At Wakulla, John Reuben Vallentyne of Cornell University complained that this question had been neglected ever since Miller's experiment.[5]

The second big problem with the primordial soup hypothesis concerned the specifics of Miller's famous experiment. By the 1970s it seemed likely, and today it is virtually certain, that his simulated primordial atmosphere was wrong.[6]

As we saw in chapter 3, Oparin argued that the early atmosphere was strongly reducing, meaning it was rich in gases like methane and ammonia that tended to donate electrons to other chemicals. Urey supported this, reasoning that gas giant planets like Jupiter are rich in these chemicals, so the early Earth was probably similar – despite being smaller and rockier. The only major difference, Urey thought, was that Earth quickly lost almost all its hydrogen, whereas the gas giants kept theirs. That's because hydrogen is the lightest chemical, so only big planets with a strong gravitational pull can hold onto it.

However, the idea that Earth once had a reducing atmosphere was already being questioned before Miller did his experiment. In a 1947 lecture, J. D. Bernal pointed out that methane in the air would not have lasted long, because it would have turned into carbon dioxide – which is less reactive and therefore not a strong reducing agent.[7]

Then, in the 1950s, an American geologist named William Walden Rubey reframed the question.[8] He assumed the atmosphere was mostly formed from gases released by volcanoes, and examined what comes out of them. He found volcanoes mostly release carbon dioxide and nitrogen, and concluded these were the main constituents of the early atmosphere.[9] Such an atmosphere would only be weakly reducing.

Most origin-of-life researchers ignored Rubey for many years, perhaps blinded by the success of Miller's experiment. But by the late 1970s the evidence had piled up.[10] Planetary scientists figured out how quickly Earth would have lost its hydrogen, allowing detailed calculations confirming that the young Earth's atmosphere was only weakly reducing.[11] By the early 1990s, most accepted that the early atmosphere was mostly nitrogen and carbon dioxide, and was only weakly reducing.[12]

The argument is not quite settled, as some researchers still support a more reducing atmosphere.[13] They argue that asteroid impacts could have released significant volumes of methane. But they have not convinced their colleagues that this made a meaningful difference. The crucial point, as Bernal spotted in the 1940s, is that reducing gases like methane and ammonia are unstable in sunlight and break down into other chemicals. Jim Kasting has calculated that even a large volume of methane would be entirely converted to carbon dioxide within thirty million years.[14] Life would have been in a race against time, needing to form before the reducing chemicals were destroyed.

The discovery that the primordial atmosphere was only weakly reducing effectively invalidated Miller's original experiment and many of the follow-ups. Any primordial synthesis experiment that relied on a strongly reducing atmosphere was irrelevant, because that was not what the atmosphere was like. Years of work were reduced to chemical curiosities that revealed nothing about life's origins. Miller's experiment remains totemic, but not because it was correct. Its importance is symbolic: a vivid demonstration that the chemicals of life could be formed by natural means. But Urey had been wrong: the synthesis Miller used was not the one nature chose.

It was much harder to obtain biological chemicals from a weakly reducing atmosphere. Miller tried in the 1970s, in an experiment simulating a neutral atmosphere that was neither reducing nor the opposite ('oxidising'), so had no tendency to either donate or gather electrons. He got little or no amino acid. Decades later his former student Jeffrey Bada had a postdoc repeat the experiment.[15] They found acid formed in the mixture and stopped the amino acids forming. When this was corrected by adding limestone, the reaction did work, but it only made tiny quantities of amino acid.

The rude awakening for the Oparin–Haldane–Miller hypothesis finally came in 1986. The Space Race was over, having been firmly won by the USA, and the USSR was on the back foot and only a few years from collapse. Perhaps in this changing climate US researchers felt less pressure to pretend that they understood how life began.

Robert Shapiro duly slated the entire field in his book *Origins*.[16] Shapiro was logical, clear and vicious. He spent much of the book explicitly stating assumptions that had until then been implicit – exposing their vacuity in the process. In particular, Shapiro tore down the idea that the primordial soup hypothesis could be rescued with only minor fiddles, and argued for a wholesale rethink. His students nicknamed him 'Doctor No', after his monosyllabic response whenever anyone proposed one of the existing hypotheses. Many researchers had already spotted the problems he highlighted, but Shapiro pointed them out in public.

One of Shapiro's key points, echoed by Cairns-Smith in *Seven Clues*, was that researchers had relied too much on the laws of chance. Even if a particular chemical reaction or process was unlikely to work, it was said, the Earth was big and had existed for billions of years. There was time for shockingly implausible things to happen, especially if they only needed to happen once. This argument seemed appealing, but closer inspection revealed that in many cases the odds were so astronomical that having a planet to play with did not help. Instead, it would be necessary to find processes that worked easily and repeatably.

With this in mind, researchers soon alighted on a central new idea, which they thought so obviously true that they rarely stated it explicitly. The notion was that living cells as we find them today are far too complicated to have arisen all at once. Not only do they have a lot of moving parts, everything depends on everything else. For instance, protein can only be made using RNA, which has to be copied from DNA, but in order to make and maintain the RNA and DNA, the cell needs proteins (see chapter 4).

The only apparent way out of this chicken-and-egg paradox was to postulate that one part of the cell arose before all the others. In other words, one of the components formed on the young Earth and survived on its own. The other bits assembled around it later.

Against this backdrop, four questions arose, each of which led to bitter arguments that would continue for decades.

The first question was, which bit of the cell came first? This is not just an argument about a chemical process: it gets to the very nature of life, because whatever component of life arose first is arguably its defining feature. Experiments like Miller's offered little guidance: they only addressed the origin of the key chemicals, not how they self-assembled into living organisms, which was evidently the tricky bit.

There are three main suggestions for which bit of life came first. Perhaps the most familiar idea is that life began with a single gene: a molecule of DNA or something like it that carried a small amount of information, and which made copies of itself. Genes, which can be altered and thus give rise to evolution, are in this reading the fundamental basis of life.

Set against this is the argument that life must first have had a way to sustain itself: a source of energy. Living organisms survive by carrying out complicated sequences of chemical reactions, known as metabolism, so this hypothesis became known as 'metabolism first'. After all, how could you have a gene – a complex molecule – without a mechanism for building it? There had to be some chemical process by which the first organism obtained energy from its surroundings, before anything else could happen.

Finally, a third group became convinced that life began with some kind of container in which all the other things like genes and enzymes could ultimately abide. This first compartment was not necessarily made of the same chemicals as modern cells, but it did the same basic job of holding everything together.

This three-way disagreement was not altogether new. Oparin and Haldane's ideas had been lumped together, but they were not identical. Both saw cell-like compartments as essential from the start, but Oparin emphasised metabolism while Haldane focused on genes. For years this was treated as a minor detail, to be sorted out later. It was only once the problems with the Miller experiment became clear that the disagreement became pressing and, ultimately, divisive.

It is worth noting that all three ideas can be stripped back to a replicating entity *of some sort*. Whether it was a gene, a set of proteins carrying out chemical reactions, or a cell-like container, the first proto-organism must have been able to copy itself. Otherwise, how would life have spread and evolved greater complexity? People disagreed about the identity of the primordial replicator, but the idea that life was about copying was omnipresent. It is even there in Cairns-Smith's clay crystals, which supposedly copied themselves.

The disagreement about which aspect of life came first emerged only gradually. Like a marriage going wrong, at first people didn't realise the problem existed. But by 1971 it was explicit, at least in academic circles. In that year, two researchers set out competing visions for life's origin. Both studies were theoretical, relying on mathematics and first-principles reasoning rather than experiments. But in them the clash between genetics-first and metabolism-first can be seen clearly. Both ideas were seminal and will resonate through the rest of the story.

The first into print was an American theoretical biologist named Stuart Kauffman. He focused on networks of chemical reactions that he dubbed 'autocatalytic sets'. If something is autocatalytic, it can make copies of itself. DNA might be thought of as autocatalytic, because of the base-pairing discovered by Watson and Crick that allows accurate copies to be made, but this only happens with help from enzymes: DNA on its own is not autocatalytic. Kauffman doubted an individual molecule ever could be, so instead he imagined a set of molecules in which A creates B, B creates C, and so on until something creates A. At this point there are two As, both able to create Bs, and so the system gradually copies itself.

Crucially, autocatalytic sets should arise from random mixtures of chemicals. Kauffman and others have calculated that the chance of an autocatalytic set existing in a mixture rises dramatically as the number of different chemicals present increases.[17] Above a certain threshold, it is virtually certain that one will pop up.

Kauffman's 1971 paper made extensive reference to 'genes', but he quickly ditched this terminology because the molecules he was describing were not behaving like genes.[18] Instead, the molecules were harvesting simple chemicals from their surroundings and using them to make copies of one another.[19] This was a metabolism-first version of life's origin, one shorn of genetic information and instead focused on taking in 'food' and using it to build and maintain a 'body' (in this case a defined set of proteins). Kauffman came to view his autocatalytic sets as 'primitive connected metabolisms'.[20]

In October 1971, a German chemist named Manfred Eigen outlined a seemingly similar idea that, on closer inspection, falls squarely into the genetics-first camp.[21] Eigen was forty-four, with a receding hairline, high forehead and almost aquiline features. Four years earlier he had shared the Nobel Prize in Chemistry for studies of fast chemical reactions.[22] Now, aided by his graduate student Peter Schuster, he turned his attention to life's origin.

Like Kauffman, Eigen imagined collections of organic

molecules bumping together in the primordial soup. Also like Kauffman, he saw that a collection of amino acids and proteins might form an autocatalytic set (although he didn't use that term). But Eigen spotted a limitation. A new protein might arise that was highly useful, but there would be no way to ensure that copies were made, so the innovation would probably be lost.

So instead Eigen envisioned a more complex set of molecules, containing both nucleic acids to give instructions and proteins to act as catalysts. Each nucleic acid would create a protein, which in turn created a copy of the nucleic acid: this bit is a simple autocatalytic set. However, each protein/nucleic acid set would also make a second nucleic acid, which in turn created its own autocatalytic set. This would continue until the starter nucleic acid was made again. Eigen called the system a 'hypercycle', because it included several cycles of chemical reactions, feeding off each other.

The hypercycle was a special kind of autocatalytic set.[23] Unlike the others, Eigen argued, it could evolve, because changes in the nucleic acids could be passed on. Eigen thought multiple hypercycles would arise in the same patch of soup and compete for resources until only one kind survived.

The hypercycle was a brilliant feat of imagination, rigorously described. It offers a compelling vision of a genetics-first origin of life, in which the first nucleic acids take control of a ragtag bunch of proteins and turn them into an organised unit.

The problem is that a hypercycle does not simply rely on a few nucleic acids and simple proteins. It needs ribosomes and transfer RNAs to convert the information on its nucleic acids into new proteins. And ribosomes, it was already clear, are ferociously complex molecular machines honed by billions of years of evolution. Hypercycles could only have existed at life's origin if a crude version of the ribosome had arisen, and no such molecule had been found or created. It was the chicken-and-egg paradox all over again.

The abstract concepts of hypercycles, autocatalytic sets and self-replicators soon became caught up in the argument between genetics-first, metabolism-first and compartmentalisation-first, which has continued for fifty years. The supporters of each idea devised ingenious experiments and found what seemed like compelling evidence to support their hypotheses. These stories are the focus of Part Three.

So much for the first disagreement, on the nature of the first life. The second dispute, which seems identical but is actually distinct, was: what was the first 'biological' molecule? Was it nucleic acids like DNA and RNA, which can store genetic information? Was it perhaps proteins, which can control and accelerate chemical reactions, and form structures, and might possibly be persuaded to carry genetic information as well? Or might it have been lipids, the fat-like molecules that make up the outer walls of cells?

To some extent, the position someone took on the first organic chemical mapped onto their view of the first life process: a researcher who believed genes came first would probably argue that nucleic acids preceded the other molecules. But the fit was far from perfect, and researchers have argued almost any combination imaginable. In particular, as we'll see in chapter 7, Sidney Fox and his colleagues argued that proteins came first, and that they were the basis not only for the first enzymes but for the first cell boundaries as well.

This second disagreement arose for a good reason: life as we find it today might mislead us about life as it once was. For example, just because modern organisms do not use proteins to carry genetic information does not mean that proteins are incapable of doing so. Indeed, if proteins arose before nucleic acids, the first life may have used them in this way, even if they were unreliable. Carl Sagan made this point at the Wakulla conference: 'Maybe there are polynucleotides which are weakly catalytic; maybe there are polypeptides which are weakly self-replicating. We should try to find out.'[24]

On top of this, there was a third disagreement: where on Earth did life originate? Many researchers followed Oparin and Haldane's lead and argued that life began in the sea. For palaeontologists this idea seemed inescapable, because the oldest animals in the fossil record are all marine creatures like trilobites. Animal and plant life only ventured onto land later, within the last half-billion years. The sea was seen as a stable environment in which simple, vulnerable forms of life could survive. But where in the sea? In the darkest depths or the shallows? In open water or on the sea floor? Since the 1970s, many researchers have become fascinated by hydrothermal vents, where hot water from beneath the seabed is pumped up into the cool ocean above. As we'll see in chapter 11, the idea that vents are the cradle of life has attracted fervent support.

Alternatively, maybe life arose on land. That is, assuming there was any: some geologists have argued that at first the oceans were too deep for any land to break the surface. Ironically, when scientists argue that life began on land, they often still place it in water. But instead of a vast ocean the focus is on a small pond or lake. In some scenarios this pond is full of rainwater: either an ordinary hole in the ground or something a bit special like a meteorite crater. In others it is more chemically exotic, something like the hot springs and mudpots in Yellowstone National Park, which are rich in chemicals rising from underground magma chambers.

Even the idea that life began in water is not universally accepted. Chemists have a major problem with it, because on their own the molecules of life break down in water. Shapiro emphasised that water tears biological molecules apart, and the fact that our cells have evolved mechanisms to repair this damage – mechanisms that would not have existed on the early Earth. A popular solution is to imagine life forming in a pond or pool that periodically dried out in the sun.

In truth, there are more proposed locations for life's origin than are worth discussing. Researchers have argued that life arose in the meltwater trickling between slabs of ice near the frozen poles; between the sheets of a mineral called mica;[25] hundreds of metres underground in solid rock;[26] in radioactive water from buried uranium deposits;[27] and even in parched deserts.[28] However, these are all minority views. Most researchers focus either on the sea or on bodies of water on land.

The fourth and thankfully final disagreement is about what we can learn from modern life.* It needs a bit of unpacking.

* Well, there is one other debate: what is the best way to approach the question of life's origin? Researchers like Miller did 'bottom-up' research: they tried to make the components of life from simple chemicals and build up a living cell. But others espoused a 'top-down' approach: examining living organisms and working backwards to see what the first forms of life were like. Which is better? The simple answer is that the two approaches complement each other. But some academics do love a pointless argument, so even today this silly debate sometimes gets trotted out.

The theory of evolution tells us all organisms are related, so if we work back far enough in time we should find the Last Universal Common Ancestor (LUCA): the species from which all life today is descended. We do not know exactly when LUCA lived, but it was certainly over 3 billion years ago. That is long before any complex animals evolved, so LUCA was surely a single-celled microorganism. Reconstructing it ought to tell us something about life's origins. For instance, if we learn that LUCA was adapted to a particular range of temperatures, or to use a certain substance as food, that will narrow down the options for where life could have arisen.

There are limits to this approach. LUCA was the ancestor of all life today, but that does not mean it was the first life on Earth. Quite possibly life had been evolving for millions of years, with many species failing and dying out, before LUCA came along. In that case, LUCA was one of many competing species of microbe, and it outcompeted all the others – with the result that LUCA still has living descendants, and none of the others do. That means there is only so much we can deduce about the first life from LUCA. If a particular gene is absent from LUCA, it seems a good bet that it was also absent when life began – but if something is present in LUCA, that does not necessarily mean it was there in the first life.

Nevertheless, biologists have tried to reconstruct the family tree of all life, and thus to uncover LUCA. A crucial step was taken in 1977 by microbiologist Carl Woese (pronounced 'woes').[29] He wanted to find out how the many single-celled organisms were related: to draw a family tree of microbes.

At this time, biologists largely agreed that there were two kinds of cell, called prokaryotes and eukaryotes. Prokaryotes are small and simple cells: bacteria are prokaryotes. Eukaryotic cells are bigger and more complex. Some are single-celled, including *Amoeba*, but they also include all multicellular organisms like plants and animals – and humans. Clearly, eukaryotes are more advanced and evolved later. Prokaryotes are simpler and older.

Woese focused on prokaryotes. He traced their relationships by examining a particular bit of RNA from over 100 species. If it was similar in two species, they were closely related: if it differed, they were more distantly related. To his surprise, he found that some prokaryotes did not match their fellows. Their RNA was as different from that of other prokaryotes as it was from that of eukaryotes. That meant there were two kinds of prokaryote, which had been evolving separately for billions of years. Bacteria form one group: the other is now called the 'archaea'.[30]

The discovery of archaea meant that all life could be divided into three groups. There were archaea and bacteria, all single-celled and profoundly ancient, and then there were the more advanced eukaryotes. Every known organism belongs to one of these three groups.

Woese's discovery allowed biologists to sketch out the history of life. After life formed, it evolved for some unknown period of time until LUCA arose. Then the first big split occurred. One group of microorganisms became the archaea and another became bacteria. Genetic evidence suggests this split took place about 3.4 billion years ago, after which bacteria and archaea spread and diversified for millions of years. The eukaryotes arose much later thanks to an extraordinary event described by

Lynn Margulis:* somehow, a bacterium wound up living inside another cell[31], perhaps an archaean.†

Between them, Woese and Margulis had established something crucial. Anyone interested in the origin of life can pretty much ignore eukaryotes, and therefore all animals and plants, as they evolved late in life's history. This means we don't have to worry about the intricacies of eukaryotic cells, about sex, or about large organisms made of multiple cells. The only relevant organisms are bacteria and archaea, because they are the oldest and can tell us about LUCA.

This is where the final disagreement lies: which group is most like LUCA? Do bacteria or archaea give us the clearest glimpse of the first life? The answer may be 'neither'. The more geneticists learn about LUCA, the more it seems to have had traits from both.

It should be clear that these four disagreements were profoundly challenging. Indeed, so desperate did researchers become that, in the 1970s, some revived an idea that had long been discarded as both wrong and pointless: that life did not form on Earth, but was carried here from elsewhere in the universe.

* Yes, Carl Sagan's ex-wife.

† Giving rise to another argument: how many kinds of life are there? Bacteria and archaea are clearly separate 'domains', as these high-level groups are known. But are eukaryotes a third domain, or a subgroup of one of the others – and, if so, which one? This is one of those splendidly semantic arguments that have nothing to do with facts and everything to do with how you interpret them, so it seems unlikely to be resolved unless the supporters of one position all die of old age before their opponents do. However, for what it's worth, eukaryotes are so different to the bacteria and archaea that it seems ridiculous not to treat them as a third domain of life.

This notion is called panspermia and has a long history. The Greek philosopher Anaxagoras (510–428 BC, more or less) may have believed something along these lines. Reportedly, he argued that 'life came from the sky in the form of semen that peopled the earth'.[32] The idea returned to prominence in the 1800s, thanks in part to William Thomson (Lord Kelvin), who we met in chapter 1 arguing that the Earth was too young for evolution to have occurred.[33] Thomson discussed panspermia in an 1871 lecture, by which time spontaneous generation had been disproved and it was known that meteorites fall from space. It was at least possible, he said, that 'life originated on this earth through moss-grown fragments from the ruins of another world'.

The word 'panspermia' was coined in the early 1900s by Swedish chemist Svante Arrhenius, who was also one of the first to foresee that greenhouse gas emissions would warm the climate.[34] He set out the case for panspermia in *Worlds in the Making*.[35] Oparin's hypothesis about life's origin was over a decade away, so the idea that life formed from non-living matter struck Arrhenius as ridiculous. Instead, everyone had to accept that 'life is eternal' and 'has always been there'. This may seem bizarre to modern readers, because we are taught that the universe had a beginning in the Big Bang: after all, life cannot be eternal if the universe is not. However, when Arrhenius was writing, physicists believed the cosmos had existed for ever. Astronomers would only discover that the universe is expanding – and therefore was once extremely small – in the 1910s and 1920s.

That aside, Arrhenius had a compelling idea for how life spread. He pointed out that stars emit a solar wind, made not of air but of radiation and subatomic particles. This could have propelled microorganisms through space. Most would 'meet death in the cold infinite space of the universe', but it would be enough if a handful survived. This was a grand vision of lonely journeys spanning millennia, undertaken in darkness and intense cold, with each organism having only a faint hope of finding safe harbour. It was also not wholly implausible.

Arrhenius's idea became superfluous once scenarios like Oparin's became better known. But in the 1970s, with the primordial soup hypothesis in trouble, panspermia came back. One of its most quixotic defences was authored by Francis Crick (of DNA fame) and biochemist Leslie Orgel (who we will meet in chapter 8). In 1973, they proposed 'directed panspermia'.[36] In this version, life was spread to new planets, not by natural processes, but by intelligent aliens purposely seeding the galaxy. Crick and Orgel argued that Arrhenius's hypothesis had been disproved, because organisms crossing interstellar space would be killed by radiation: a claim that was sound at the time and now seems virtually unassailable. Instead, they suggested that aliens infected Earth with life out of 'missionary zeal'.

Their underlying reasoning was that panspermia offered an answer to the seeming improbability of life getting started. If the spontaneous formation of a living organism is astronomically rare, then the only way to explain life on Earth is to invoke the astronomical scale. If life can be carried between planets and even between solar systems, then that one-in-a-trillion event only has to have happened once in our galaxy.

Crick and Orgel do not seem to have been entirely serious about directed panspermia. Certainly, neither invested further effort in it. Instead, they may have been frustrated by the lack of progress on the origin of life. It is tempting to read it as a challenge to colleagues: 'look at the drivel we'll have to invoke if you don't come up with better ideas'.

However, if Crick and Orgel were tongue-in-cheek, Fred Hoyle was in deadly earnest. He was an English astronomer who built a prominent career by being alternately extremely right and noisily wrong.[37] He helped show that almost all the chemical elements were created in the hearts of stars. Only the three lightest elements – hydrogen, helium and lithium – were made immediately after the Big Bang. Everything else, from carbon to uranium, was made inside stars. This is the meaning of Carl Sagan's famous line, 'We are made of starstuff.'[38] However, Hoyle also opposed the Big Bang theory, instead pushing the now-disproved idea that the cosmos had existed for ever in a 'steady state'.

Hoyle's other act of scientific blasphemy was to reject the idea that life formed on Earth, and instead promote panspermia. His main collaborator was Chandra Wickramasinghe, a Sri Lankan astronomer who was his student in the 1960s, and who has continued pushing panspermia since Hoyle's death in 2001.[39] At first, the pair studied interstellar dust grains and comets. They argued that carbon-based chemicals might be present, including substances found in life, like sugars. This was controversial, but as we will see in chapters 12 and 14, they were correct. Many of the chemicals of life can be found in space.

However, Hoyle and Wickramasinghe then went a lot further. From the 1970s onwards, they argued that living cells came to Earth from space. Furthermore, they claimed that life is still raining down on Earth, and that this is the source of major disease outbreaks, including the 1918 Spanish flu and the AIDS epidemic. They offered no evidence for this.

On top of that, Hoyle and Wickramasinghe argued that life from space has integrated itself into existing organisms, triggering some of the key transitions in evolutionary history, including the origin of birds. Lacking any evidence for this, they instead tried to discredit the theory of evolution. In 1986 they argued that a fossil of *Archaeopteryx*, one of the first dinosaurs to have wings and therefore a key step on the road to birds, was a fake. This was patently ridiculous.[40] They had no evidence for the forgery, and there are multiple *Archaeopteryx* fossils and plenty of similar dino-birds, so discrediting one fossil would mean little. The claim has since become a talking point for creationists, a frankly embarrassing situation for which Hoyle and Wickramasinghe must take their share of the blame.

Compounding the offence, Wickramasinghe testified at a 1981 trial over whether Arkansas schoolchildren should be taught creationism as well as evolution – on the side of the creationists.[41] Wickramasinghe presented his arguments for panspermia and argued that evolution is 'woefully inadequate' to explain life on Earth.[42]

In recent years, Wickramasinghe's claims have only become more peculiar. In 2013 he claimed that a meteorite that had landed in Sri Lanka contained the fossilised remains of single-celled organisms called diatoms.[43] This, he argued, was proof that microbes could travel through space in meteorites. However, the diatoms were probably not fossilised, but recently dead – suggesting they were the result of contamination. Worse, it was not clear the rock was a meteorite: it was an irregular lump, while meteorites are normally rounded from their passage through the atmosphere.[44] Biologist and blogger P. Z. Myers mocked the study as 'Diatoms . . . iiiiin spaaaaaaaaaaace!'[45]

Wickramasinghe refused to accept that there were any issues.

Instead, he claimed to have identified a fragment of a diatom over twenty-five kilometres up in the air.[46] It had been captured by a high-altitude balloon. Wickramasinghe argued that the diatom must have come from space, because there was no obvious way for it to have got so high up. However, only one partial diatom was found, so the odds against the find are not so astronomical that we need to invoke an extraterrestrial origin.

Wickramasinghe shows no sign of stopping. In 2018 he argued, for no clearly defined reason, that octopuses could not have achieved their remarkable intelligence through evolution.* Compared to related animals, they have a lot of new genes, which he claimed are 'extraterrestrial imports' carried in on 'cryopreserved Squid and/or Octopus eggs'.[47] This, he wrote, offered 'a parsimonious cosmic explanation' for octopuses.† In fact, the octopus fossil record is pretty good, and octopus eggs are even less likely to survive in space than microorganisms. It is hard to imagine a sillier idea.

Clearly, Chandra Wickramasinghe is not a reliable source of information on panspermia, or indeed biology. But that does not mean that panspermia is untrue. Setting aside his outlandish claims, does the idea hold any water? Most scientists argue that it does not, for two reasons: one philosophical, the other evidence-based.

* Fans of the writings of H. P. Lovecraft may suspect that this is an elaborate ploy to infuriate the dread god Cthulhu so that he finally destroys humanity.

† The term 'parsimonious' crops up four times in Wickramasinghe's octopus paper. The thing is, although he keeps using the word, I do not think it means what he thinks it means.

The philosophical problem with panspermia is that it is cheating. Instead of devising a process by which life could have begun on Earth and testing it, advocates of panspermia evade the challenge by saying that the formation of life was extremely unlikely, only happened once in the galaxy, and cannot be repeated. This is not an explanation, but an admission of defeat. In other words, panspermia has a similar problem to 'God did it'. If we invoke a god to explain life, it raises the question of where the god came from. The problem has been pushed back a step, not solved. Similarly, panspermia pushes the problem of the origin of life to another world.

Shapiro saw this clearly. 'We have hardly examined the opportunities here,' he wrote, 'there is no need to turn elsewhere.' Of all people, Karl Marx made the same point in 1875, four years after Thomson's pro-panspermia lecture. In a letter, he mocked 'the absurd doctrine that the germs of terrestrial life . . . were brought down here by aerolites'.[48] His objection was simple: 'I detest the kind of explanation which solves a problem by consigning it to some other locality.'

Meanwhile, the evidence-based problem is simple. We have not found any evidence of life in space, and earthly organisms cannot survive there for long. If bacteria were raining on Earth from space, we should be able to find some up there, and we haven't. Furthermore, astronauts have repeatedly exposed microorganisms to the vacuum of space. While some endure longer than others, sometimes over a year, they all succumb in the end.[49] This tells us that microbes may survive long enough to travel within our Solar System, perhaps between Earth and Mars, but interstellar travel is unlikely. This removes the one apparent benefit of panspermia: if life could only reach Earth from the handful of worlds in our Solar System, that doesn't make much difference to the supposedly huge odds against it forming.

For this reason, most origins researchers view panspermia as a distraction. Better, they say, to face the difficulties head-on.

This chapter has thrown up a lot of questions, so let's summarise. By the 1980s, origins researchers disagreed about four major issues: which of life's essential functions came first; which of life's basic molecules formed first; where on Earth this happened; and which organisms held the clues to these questions. There was also the niggling issue of what the early Earth was like: which gases were present in the air and whether there was land.

As we've seen, the various questions were not entirely independent in people's minds. If you knew what a researcher thought about one of them, you could make a good guess at their view of the others. As a result, this tangle of questions became resolved into four main schools of thought. Each of these new ideas attracted devoted supporters, some of whom became virtually fanatical about their chosen hypothesis over the coming years. Where once there was collegial disagreement, there was now infighting. But there was also a torrent of big ideas and ingenious experiments, which have shed a lot of light on how life began. In Part Three we will explore each of the ideas in turn. While it seems unlikely that any of them is the full explanation of life's origin, each nevertheless offers crucial clues to the answer.

PART THREE

Scattered, Divided, Leaderless

'But I am very poorly today & very stupid & hate everybody & everything. One lives only to make blunders.'

Charles Darwin, in a letter to geologist Charles Lyell, dated 1 October 1861[1]

Chapter 7

The Other Long Molecule

Two major things happened in France in 1789. The first was the French Revolution, during which the monarchy was overthrown and the king executed by guillotine – ultimately leading to the dictatorship of Napoleon Bonaparte and decades of war. The second was arguably more momentous, but it tends to be overlooked.

In that year the chemist Antoine François, comte de Fourcroy was studying the chemical makeup of living organisms. The previous year he had identified three classes of matter that could be found in animals. There was 'gelatin' from skin; 'albumin' from milk and eggs; and 'fibrin' from muscles. In 1789 he reported a similar study of wheat plants, in which he identified both albumin and a fourth substance called 'gluten'. It was clear that these mysterious substances were somehow central to life.

Fourcroy was the first scientist to identify proteins, one of the four types of molecule that underlie all life on Earth.[1] His discovery was foundational, but it would soon be overshadowed.

As the Revolution took hold of France, so did paranoid

suspicion, and soon people were being guillotined without fair trials, or sometimes any trial at all. One such victim was Fourcroy's close colleague Antoine Lavoisier, who arguably did more than anyone else to found the science of chemistry. Unfortunately for Lavoisier, he was both an aristocrat and a tax collector, and was guillotined in 1794. An early biography of Lavoisier blamed Fourcroy, accusing him of failing to stand up for his friend.[2] Later scholars have disputed this, and in the absence of hard evidence of collaboration this seems only fair. The situation had spun out of control, any protest could be taken as a counter-revolutionary act worthy of execution in itself, and Lavoisier's background made him difficult to save.

Perhaps partly because of the chaos unleashed by the Revolution, Fourcroy's work would not be picked up for decades. Finally, in 1838, the Dutch chemist Gerardus Johannes Mulder took the next step. He took fibrin, albumin and gelatin and broke them down into their individual elements.[3] All contained carbon, hydrogen, oxygen, nitrogen, potassium and sulphur. Crucially, the ratio of these atoms in each molecule was fairly consistent: 'in the egg albumin there is always 1 atom of sulphur to 1 atom of phosphorus'. Mulder took this to mean that they were all essentially the same type of molecule. His colleague Jöns Jacob Berzelius coined the word protein to describe these substances.[4]

Proteins were evidently gigantic by molecular standards. Mulder's analysis suggested that each molecule contained over 1200 atoms. At the time, many chemists thought that such large molecules could not exist, because they would fall to bits at the slightest nudge, like a house of cards. This finding would set the stage for the realisation that life relies on large 'macromolecules' like DNA.

It took the rest of the century to figure out what proteins were made of. The key was the discovery of simpler molecules called amino acids. The first had been isolated from asparagus back in 1806, and thus named 'asparagine'. Others followed throughout the 1800s and early 1900s.

Amino acids all have the same structure, except for one part of the molecule. There is a central carbon atom surrounded by four things. The first is an amine: a nitrogen atom with two hydrogens dangling from it. The second is a carboxyl: another carbon, this one holding two oxygens and a hydrogen. The third is another hydrogen and the fourth is a *something* – the bit that varies.

Life on Earth uses twenty amino acids, although there are many more. The simplest is glycine, in which the *something* is merely another hydrogen atom. In contrast, in asparagine the *something* contains eight atoms, about as big as an entire glycine. Others are similarly intricate.

Amino acids were sometimes released when proteins broke down, so there was some relationship between the two. Then on one dramatic day in 1902, Franz Hofmeister and Emil Fischer both gave presentations at a scientific conference, suggesting the same idea. Hofmeister was a reclusive type who repeatedly turned down prestigious university positions. Fischer was a devoted experimenter who had previously suffered mercury poisoning.

Hofmeister was on in the morning, Fischer in the afternoon. Both argued that proteins are long chains of amino acids, linked together. This explained why proteins were so similar and yet so different: their atomic makeup was similar, but different proteins had varying properties. This now made sense, because the different amino acids could be strung together in any order, creating wildly distinct molecules out of the same small set of building blocks.* Anyone who has built both a dinosaur and a rocket out of the same Lego set can imagine the degree of versatility these molecules have.

Almost half a century later, the British biochemist Frederick Sanger became the first to figure out the detailed makeup of a protein. He determined the sequence of amino acids that makes up insulin, the production of which is affected in diabetes. In a series of papers between 1949 and 1952 that won him the first of two Nobel Prizes, Sanger showed that insulin contains fifty-one amino acids, in two chains that are linked at two points.[5] Since then hundreds of proteins have been sequenced.

Chemists have also figured out the three-dimensional structures of proteins. While early protein researchers treated them as spheres or ovoids, we now know that they can be startlingly intricate. The first structure to be determined was that of myoglobin, a protein found in muscles, in 1958.[6] The X-ray images were not precise enough to show the positions of individual atoms, but they revealed the overall twisting of the amino acid chain. The resulting image looked like a strand of Play-Doh that had been tangled and knotted. Three years later, the picture had been sharpened and each atom was pinpointed. Nowadays thousands of proteins have been imaged to this standard.

We now know that proteins perform a staggering variety of roles in living organisms. They make up the internal scaffolding of cells (the 'cytoskeleton'). Some are pumps that straddle

* Just to make things more complicated, short chains of amino acids are called 'peptides'. The term 'protein' is reserved for longer ones, typically with more than fifty amino acids. Throughout this book, for the sake of simplicity, we will call chains of amino acids 'proteins' no matter how short they are.

the outer walls of cells and shuttle small molecules in and out, ensuring that the cell obtains essential nutrients but doesn't get poisoned. Others are sensors, like the rhodopsin in our retinas, which responds to light and allows us to see.

Perhaps most impressively, they can form enzymes: the molecular machines that drive the chemical reactions taking place in living cells. Enzymes are 'catalysts', meaning they are not permanently changed by the chemical reactions they control, so each enzyme can be used over and over.*

It was not immediately obvious that enzymes were made of protein, but the American chemist James Sumner showed that they were in 1926. He spent nine years crystallising an enzyme called urease and demonstrating that it was a protein.[7] Sumner worked largely alone, with little money and with only one arm: he had lost his left arm in a shooting accident as a young boy. This was especially inconvenient, as he was left-handed.

The structures of enzymes are beyond the imagination of human engineers. Often several amino acid chains are coiled around each other in intricate spirals and folds, creating a perfectly shaped 'active site' where the enzyme can clamp onto the molecule it is working on. Emil Fischer, who helped determine what proteins are made of, described this 'lock and key' mechanism in 1894.[8] A living cell is a seething mass of production lines powered by hundreds of different enzymes, all churning out useful chemicals in less time than it takes to blink.

Not all of this was clear to the Canadian biochemist Archibald

* The word 'catalysis' was coined by Berzelius in 1836, not long before he came up with 'protein'.

Macallum in 1908. He knew that proteins are chains of amino acids, but the full scope of their structure and behaviour was yet to be realised, and Sumner's proof that enzymes were made of protein was years away. Nevertheless, it was apparent that proteins were central to life, so Macallum argued that they might also have been the first form of life.[9] He seems to have been the first person to suggest this in a published paper – although Darwin beat him to it in his letter.

Macallum was evidently inspired by Hofmeister and Fischer's discovery, which was then just six years old. After outlining the evidence that proteins are made of amino acids, Macallum argued that this offered an explanation for life's origin: 'If we could explain how proteins first arose without the participation of living matter in the synthetic act we might be in a position to explain the origin of living matter itself.'

He went on to discuss the simplest possible proteins, in which two amino acids are strung together. Such 'dipeptides' are common in life, and while they are simple they remain diverse: the sweetener aspartame is a dipeptide, as is carnosine, which is found in meat. Macallum reasoned that such simple molecules would be more likely to form than intricate proteins like insulin, and could still perform many of life's vital functions. This led him to a fascinating speculation: 'It is conceivable that ultramicroscopic particles consisting of but a few molecules of protein appropriately constituted could live separately and reproduce themselves by division when the particles had grown beyond a certain size.' However, Macallum's suggestion did not garner many followers, so the idea that life began with proteins languished until the 1950s – when it found a vociferous champion.

*

By the time Miller published his experiment showing that amino acids could seemingly have existed on the young Earth, Sidney Walter Fox was already established as a biochemist. Born in Los Angeles in 1912, he looked like a thickset, bespectacled version of Matt Damon. He married a Russian woman, Raia, and they had three sons who all became scientists.[10] He set up a lab at Florida State University in 1955.

Inspired by Miller's experiment, Fox wanted to take the next step: to make simple proteins from the amino acids Miller had made. When two amino acids fuse, a molecule of water is released, as the molecules shed two hydrogens and an oxygen. This led Fox to the idea that he could prod amino acids into linking up by driving off the water: in other words, by simply heating the amino acids.[11]

At first this did not work at all and Fox ended up with a black, tarry mess. But then he realised that two amino acids are particularly common in proteins: aspartic acid and glutamic acid. So in 1958 his team performed a series of dramatic experiments.[12] They found that aspartic acid would link up with any other amino acid if it was gently heated. This led Fox to wonder if it could combine with all the other amino acids at once. 'If it could,' he wrote, 'we might have something like protein in a way so unexpectedly simple that its spontaneous occurrence on primitive Earth could be imagined without strain.'

At the time, Fox had two graduate researchers, Allen Vegotsky and Kaoru Harada, plus a technician named Donna Keith. The experiment to link up all twenty amino acids was considered hopeless, so Keith was assigned the job – the graduates didn't want to waste their time. Keith heated a mixture of amino acids loaded with aspartic acid and glutamic acid for several hours at 170°C, and obtained tiny grains of an unknown, off-white substance.

Harada offered to check the results, breaking the product back down into its constituent amino acids and performing a test to see how many were present. To everyone's surprise, the test identified fifteen or sixteen amino acids. Follow-up experiments obtained the lot.[13]

In other words, the grains Keith had made consisted of amino acids that had linked up. Unlike typical proteins, they were not in neat chains, but in all manner of shapes and combinations. Fox called them 'proteinoids', because while they were not quite proteins, they were entirely made of amino acids – just jumbled up. The results were published in 1958, with Fox and Harada as the sole authors.[14] A small footnote acknowledged 'the technical assistance of Mrs. Donna Keith'.

The following year, Fox's team produced an even more startling result: tiny globules of proteinoid less than a thousandth of a centimetre across.[15] These 'spherules', or 'microspheres' as they became known, looked like simple cells. A new lab assistant, Jean Kendrick, made them after Fox had her perform another seemingly simple experiment: boiling some proteinoid in salt water for a minute. One such run could easily make over a million microspheres, which lasted for weeks.

Fox immediately compared the microspheres to coacervates, which Oparin said had formed the first crude cells (see chapter 2). Fox soon concluded that his proteinoid microspheres were a better candidate for the primordial cell than any coacervate. They were so easy to make: one just had to heat some amino acids to make the proteinoids, plunge them into hot water, then let them dry. In 1960 Fox set out his argument, which he would pursue for the rest of his life, that proteinoid microspheres or something very like them were the answer to the question, 'How did life begin?'[16]

Fox became firmly convinced that his proteinoid microspheres were superior to coacervates after visiting Oparin in Moscow in 1969.[17] He asked to see the coacervates, but the resulting demonstration did not go well. A technician, supervised by Oparin, tried to make them in front of him and failed repeatedly. 'Her and Oparin's disappointment and embarrassment were obvious.'

In the early years, Fox was cautious about making big claims. At Wakulla in 1963 he was conspicuously circumlocutory, saying: 'I think it is rather too easy to assume – and in making this statement, I don't necessarily argue for the opposite point of view – that because the first administration of the events in the biochemical history of the cell are under the jurisdiction of DNA and RNA, or under the jurisdiction of RNA, that in the development of prebiological systems RNA and DNA preceded the protein. I believe it unnecessary and perhaps psychologically inhibitory to make this assumption at this time.'[18] Few bets have been more comprehensively hedged.

But over the next twenty years Fox became much more forthright. In part this was because his ideas faced a backlash, so he fought back. Alan Schwartz, who was his PhD student from 1960 to 1965, recalls Fox being 'vehemently attacked' by other chemists at Florida State University after giving a lecture. One colleague, DeLos DeTar, said something to the effect of 'Only God can make a protein'.

That was an emotional critique, but Fox was also fending off more considered arguments. At Wakulla, Carl Sagan slammed his ideas as 'geologically implausible', complaining that, to make the microspheres, the starting materials had first to go through a wide range of temperatures, then had to be variously wetted and dried.[19] 'This sequence may be convenient in the laboratory; whether it occurs frequently in nature is less clear,' Sagan argued. This was the same critique that was bedevilling chemists like Ponnamperuma: nice experiment, but would it happen in the real world?

In response, Fox spent years amassing evidence that the microspheres did not just look like living cells: they behaved like them. By 1965 he had evidence that they could divide into two, forming 'daughter' microspheres, just like modern cells dividing.[20] By 1980 it was clear that both the proteinoids and the microspheres were catalytically active, albeit weakly so: they could function as crude enzymes.[21] This seemed to hint that proteinoid microspheres might be able to make other crucial molecules, like nucleic acids, by speeding up the key chemical reactions.

Unfortunately, Fox often went too far, venturing into unsupported assertions. By 1988 he was claiming that his microspheres had a propensity to pair off, which he compared to sexual attraction. He even compared the random motion of the microspheres, endlessly jostled by smaller molecules, to mating dances.[22] However, single-celled organisms like bacteria do not reproduce sexually, but by simply dividing in two. Sex was invented surprisingly late in the history of life.

Fox was clearly too enthusiastic about his results, says Schwartz. 'He went on quite a limited amount of data, suggesting things in the direction he was claiming, but never completely fleshed it out in detail,' he argues. 'He did it out of enthusiasm, he wasn't trying to fool anyone. He would select certain properties and show that the products had these properties, and then jump onto something else.'

Fox's overconfidence showed in other ways. When Schwartz began his PhD in 1960, Fox suggested an ambitious project. Fresh from the success with the proteinoids, Fox proposed making nucleic acids like DNA in the equivalent way: by heating mixtures of nucleotides. Schwartz spent three years in a basement laboratory making 'black messes'. He now calls it 'totally crazy'. Fox's supervision was very light-touch: if he had paid closer attention, Schwartz might have wasted less time on a doomed project.

When Fox died in 1998, the protein-first hypothesis lost its most charismatic advocate – even if he had lost some of his credibility. Since then, according to Schwartz, 'he has been kind of neglected'. While the Miller–Urey experiment remains totemic, Fox is half-forgotten.

But the idea he promoted, that life began with protein, has never quite gone away. Robert Shapiro made a spirited argument for it in *Origins*, after (as we saw in chapter 6) dismissing virtually everything else.* Another supporter was Freeman Dyson, a physicist who, despite his myriad achievements, is best known for devising the Dyson Sphere: the idea of building an artificial sphere around a star to harness all of its energy output. Dyson proposed that 'the original living creatures were cells with a metabolic apparatus directed by protein enzymes but with no genetic apparatus'.[23]

* After building much of his book around a chicken-and-egg metaphor, Shapiro called his pro-protein chapter 'The case for the chicken'.

What's more, the idea has had some surprising successes. In 1996, a team led by M. Reza Ghadiri created a protein that could self-replicate: it could copy itself.[24] The protein was just thirty-two amino acids long and was able to join together two shorter proteins of fifteen and seventeen amino acids, each of which matched part of the original. The new copy of the original protein could make copies of itself in turn.

As we saw in chapter 6, researchers like Stuart Kauffman had identified the ability to copy oneself (by however many intermediate steps) as fundamental to life's origin. Demonstrating it in a protein was a feat of biochemistry. However, the experiment was rather artificial. The protein could only self-replicate if it was supplied with the two halves of itself – raising the question of where the two halves came from. Nevertheless, the study showed that a protein could self-replicate, a trick mostly associated with nucleic acids.

The following year, the team took things a stage further by creating a simple hypercycle: a set of molecules that contains self-replicators and can also replicate as a whole.[25] This was the kind of autocatalysis described by Manfred Eigen. The team first built a second self-replicating protein, which assembled itself from one of the same fragments as the original plus a new fragment. When all three components were mixed together, to everyone's surprise the two proteins catalysed each other's formation, as well as making more copies of themselves.

Also following in Fox's footsteps, a team led by Stefan Schiller of the Albert Ludwig University of Freiburg in Germany has described simple proteins just five amino acids long, which naturally clump together to form hollow spheres resembling cells.[26] These protocells can survive temperatures of 100°C and host long molecules in their interiors, including enzymes and ribosomes. Chemically, they are different to Fox's microspheres, in which the amino acids were stuck together at random. But their properties are eerily similar.

Clearly, the idea that life began with proteins remains an active area of research. But since Fox's death the idea has lacked a noisy, eloquent proponent, so it is often half-forgotten. And even in the 1980s, when Fox was still going strong, the idea was on the wane. A new and exciting hypothesis had come along, which would sweep many along in its wake.

Chapter 8

Rise of the Replicators

In 1986, physicist-turned-biologist Walter Gilbert published a paper in the top journal *Nature*.[1] It contained no new ideas, experiments or equations. But it did include a name: a name that encapsulated a host of tentative ideas that had been brewing for almost thirty years, crystallising them into a single, totemic hypothesis for the origin of life. That name was 'the RNA World'. It was then, and is now, one of the most well-supported proposals ever put forward for how life might have got started. It is also arguably the first modern hypothesis: unlike Oparin's and even Fox's, it was conceived in the full knowledge of the intricacy of living cells.

The RNA World is supposed to resolve one of the central conundrums of the origin of life. In all living organisms, DNA makes RNA, which then makes protein. This leads to a chicken-and-egg paradox: which came first? The answer, Gilbert argued, was RNA – because RNA is a jack-of-all-trades molecule that could perform the functions of both DNA and protein. That is, RNA could both carry genes, as DNA does, and act as an enzyme controlling the rates of chemical reactions, as proteins do.

Gilbert imagined that evolution began with RNA molecules that could 'assemble themselves from a nucleotide soup'. These RNAs would have developed 'an entire range of enzymic activities'. Later, they would have started to make simple proteins, which made for better enzymes. Finally, DNA emerged and took over the job of carrying genes. RNA, Gilbert argued, was then 'relegated to the intermediate role that it has today – no longer the centre of the stage'. But there were still clues to the primeval RNA World in modern biochemistry, for anyone who looked.

Gilbert was not the first to argue that RNA came first. The Russian geneticist Andrei Belozersky, a former student of Oparin, had reasoned in a similar way in the 1950s, and presented his ideas at the 1957 Moscow conference.[2] Carl Sagan mentioned RNA-first in 1963 at Wakulla.[3] However, the idea merely chugged away quietly in the background. Gilbert's rebranding put it centre stage.[4]

The RNA World is a specific form of the more general genes-came-first or nucleic-acids-came-first hypothesis discussed in chapter 6. Two decades before Gilbert's paper, three people set out the argument that nucleic acids came first in the history of life. Two we have already met: Carl Woese, who traced the family tree of life, and Francis Crick, who helped uncover the DNA double helix. The third was an English chemist named Leslie Orgel.[5]

Born in London in 1927, Orgel became a Fellow of the Royal Society in 1962, aged just thirty-five. He was self-assured and incisive, with little tolerance for people less intelligent than him, and a tendency to be brusque. In later years he would obsessively collect Persian saddlebags.* He lent his name to a pair of axioms about biology, dubbed 'Orgel's rules'. The second rule is the famous one: 'evolution is cleverer than you are'. It's an anti-creationist talking point: just because someone cannot see how evolution might have created something ingenious like the eye doesn't mean it couldn't; it just means that person doesn't have enough imagination.†

When Orgel was convinced he was right, he could not be budged. In 1973, researchers had formed the International Society for the Study of the Origin of Life. After Oparin's death in 1980, the society began handing out an 'Oparin Medal' to its most distinguished members. In the 1990s they decided to give it to Orgel, but he would not accept it, because he despised Oparin for his collaboration with the Soviet state. Orgel was only mollified when the society decided to have two alternating awards, only one of them named for Oparin, and in 1993 he duly accepted the Urey Medal.[6]

* Orgel also experimented with what he called 'eutectic wine'. The idea was to cool wine until the water froze but the alcohol remained liquid. At that point the water could be removed and the alcohol, which would hopefully hold all the flavour, could be stored. This way large quantities of concentrated wine could be carried around more cheaply and easily, needing only to be rehydrated when it was time to drink. Orgel somehow talked his graduate student Gerald Joyce into actually doing this for him. However, when the reconstituted wine was sampled during a camping trip, the results proved unacceptable. Fortunately, some non-eutectic wines had been brought along as controls.

† Orgel's first rule is: 'Whenever a spontaneous process is too slow or too inefficient, a protein will evolve to speed it up or make it more efficient.' Strangely, it didn't catch on as well as the second one.

Orgel was quickly drawn to the origin of life. In 1953, he was one of five researchers who drove from Oxford to Cambridge to inspect Watson and Crick's model of DNA before it was published. He subsequently moved to America and began researching life's beginnings in earnest.

Orgel, Crick and Woese made the case that nucleic acids came first in the late 1960s. The first to argue this was Crick, in a lecture given in December 1966. Orgel told him he was working on similar ideas, and the two decided to publish co-ordinated papers. These came out in 1968.[7] However, they were beaten into print by Woese, whose book *The Genetic Code* hit bookshelves in 1967.[8]

Much of Orgel's paper was devoted to arguing that proteins, despite Fox's then-growing evidence base, could not have kick-started life. His reasoning was simple: the first living molecules had to be able to make copies of themselves. Nucleic acids were set up for that, because of the way the bases on the chains could pair up (as we saw in chapter 4), whereas proteins were not. Therefore nucleic acids must have come first.

He did not get any more specific than that, writing: 'I do not feel obliged to decide whether DNA preceded RNA or *vice versa*.' But he did make a prediction that would be crucial to Gilbert's vision of the RNA World: that nucleic acids could act as enzymes if they folded up correctly.

Woese was equally cautious, focusing on how information was converted from bases in DNA to amino acids in protein, and on how this intricate process evolved. On the question of the first genetic material, he simply said it could have been 'either RNA or DNA'.

Only Crick went all-in for RNA. After pointing out that the

ribosomes that make proteins are partly built of RNA, he wondered 'if the primitive ribosome could have been made entirely of RNA'. Then he went further: 'It is thus not impossible to imagine that the primitive machinery had no protein at all and consisted entirely of RNA.'

Over the following decades, Orgel devoted much of his time to demonstrating that nucleic acids could have come first. In 1973, he tried to show that nucleic acids could have formed spontaneously.[9] In living cells, enzymes called replicases are responsible for stitching nucleotides together into nucleic acids. Without them, the nucleotides just sit there. Orgel supplied energy by adding small chains of phosphates, which 'activated' the nucleotides and allowed them to link up. The phosphate chains were not a random choice: they were a hugely simplified version of adenosine triphosphate (ATP): a molecule that is central to all life, as cells use it to store energy (see chapter 3).

Orgel also tried to persuade nucleic acids to self-replicate: that is, to get a second strand to assemble itself using an existing one as a template. But he struggled. By 1976 the best he had managed was to show that short chains formed more quickly if a template was present.[10] It was neat, but the experiment used modified nucleotides rather than genuine ones.*

So things puttered along, until in November 1982 a shock discovery from an unexpected source thrust RNA itself firmly into the spotlight.

* Specifically, 2'-amino-2'-deoxynucleotides. The apostrophes are pronounced 'prime', so these chemicals are called 'two prime amino two prime deoxynucleotides', which sounds like something from bad *Transformers* fan fiction.

*

The key figure was Thomas Cech, an American biochemist of Czech extraction.[11] As a child Cech was fascinated by minerals, and often knocked on professors' doors in his local university, asking to see crystal structures. But as he entered his twenties his attention shifted to nucleic acids. By 1978, he and his wife Carol had faculty positions at the University of Colorado Boulder.

Cech began studying the genes of a single-celled organism called *Tetrahymena thermophila*, which looks like a microscopic peanut covered with hairs.* His team focused on one gene, which coded for the RNA in ribosomes. When the gene was copied onto a strand of messenger RNA, part of it kept becoming detached. It was as if something was cutting it out with scissors, always at the same points. Cech assumed an enzyme was responsible, but no enzyme could be found. Even when they subjected the system to 'multiple forms of abuse' to obliterate any protein, the RNA still got cut out.[12]

The solution was simple and astonishing: the RNA fragment was cutting itself out. It was able to fold up and act as an enzyme, snipping away at the chemical bonds on either side of it. This was the first RNA enzyme ever identified.[13] During a 'relatively subdued celebration' in the lab, Cech's team discussed possible names for their discovery. They settled on 'ribozyme': an enzyme made of ribonucleic acid (i.e. RNA).

* *T. thermophila* is gloriously bizarre. It reproduces sexually, but unlike most familiar organisms there are not two sexes, male and female. Instead there are seven, known as 'mating types'. A *T. thermophila* of one mating type can have sex with any different mating type, leading to twenty-one possible combinations.

While Crick and Orgel had suggested that RNA-based enzymes played a role in the first life, nobody had thought they might still exist: the assumption had been that protein enzymes had replaced them. The discovery of the first ribozyme was an instant turbo-boost to the idea that life began with RNA.

A year later, in 1983, a second ribozyme was unveiled: an enzyme called ribonuclease P, which cut up RNA into smaller pieces. It contained chains of RNA and protein tangled together into a shape resembling a deformed mushroom. Sidney Altman at Yale University found that the RNA could carry out chemical reactions on its own.[14] Unlike Cech's ribozyme, which performed an operation on itself, this ribozyme was the same at the end as when it started: 'a true catalyst'.

It was the discovery of ribozymes in particular that prompted Walter Gilbert's article proclaiming the RNA World. Gilbert took a loosely related set of suggestions and findings, and hammered them together into a single monolithic hypothesis. The origin of life became exciting again, after years stuck in a soupy quagmire.

In particular, the quest for a self-replicating RNA – an essential element of the RNA World hypothesis – now kicked into overdrive. Cech boosted it in 1988, when his team showed that the *Tetrahymena* ribozyme could add nucleotides onto a primer molecule, creating nucleic acids ten to eleven nucleotides long.[15] Cech, once circumspect about his work's relevance to the origin of life, was now full-throated, writing that the finding 'supports theories of prebiotic RNA self-replication'.

This inspired a biologist named Jack Szostak (pronounced 'Shostak') to see if he could make a better version. His team* showed that the ribozyme could link together several short nucleic acids into one long one, if they were first positioned on a template strand.[16]

This was promising, but two years later he went one better. A student named David Bartel created a pool of random RNAs and tested to see which ones showed any catalytic activity. He pulled out those that did, randomly tweaked them, and tested them again. After ten rounds of this, Bartel had created a ribozyme that could stick together two small pieces of RNA, much as modern protein enzymes do.[17]

However, the ribozyme was not making a copy of itself, but rather a distinct RNA, so self-replication was still out of reach. The previous year, Orgel had confidently written that it 'will be achieved in the next decade'.[18] But he was wrong. Researchers have inched closer to an RNA that can copy itself, but they have not got there.

Bartel made a big leap in 2001 by creating a new ribozyme called R18.[19] It could add new nucleotides onto a piece of RNA, by accurately following a template. This was true copying, but there was a catch. R18 was 189 nucleotides long, and could only reliably add eleven nucleotides to an RNA. That is, it could only copy 6 per cent of its own length. By definition, any self-replicator had to copy 100 per cent of its length.

Arguably the best attempt at a self-replicator came from Philipp Holliger of the Laboratory of Molecular Biology in

* Szostak's co-author on this study was PhD student Jennifer Doudna, who has since become one of the most prominent scientists in the world. She is one of the pioneers of a gene-editing technology called CRISPR which promises to revolutionise biotechnology and medicine.

Cambridge, UK. In 2011, his team evolved a new version of R18, which rejoiced in the name tC19Z.[20] This could copy sequences ninety-five nucleotides long, 48 per cent of its length. This was better, but still short of what is needed.

Haldane had foreseen this problem. At the 1963 Wakulla conference, he was asked by biochemist John Buchanan whether RNA could be duplicated without a protein enzyme.[21] Haldane couldn't see how this could be achieved. 'I think it might be,' said Buchanan, to which Haldane replied simply: 'Good luck. I hope so.'

It may be that someone will eventually create an RNA that can copy itself. But the fact it has proved so difficult is itself cause for concern. If such a ribozyme does not emerge in a clean lab when someone is pushing for it, what are the chances that it emerged in the messy conditions of the primordial Earth?

This problem has led a lot of researchers to suspect that there must be something wrong with the idea of a self-replicating RNA, and thus with Gilbert's vision of the RNA World.

But perhaps the idea of a self-replicating RNA could be tweaked, rather than abandoned. One popular suggestion was that the RNA was not floating around in primordial soup but was instead on a surface, perhaps a mineral such as clay. This was not the same as Cairns-Smith's idea from chapter 5: he had envisioned clay as evolving in its own right, whereas now the clay was a kind of support system for the evolving RNA.

The new use of clay was promoted by an American chemist named James 'Jim' Ferris.[22] He first got involved in origin-of-life research in the 1960s, working with Orgel. By the late 1980s he was well into his fifties and clearly in his scientific prime. As the

RNA World came to prominence, Ferris turned his attention to a kind of clay called montmorillonite.* It forms when volcanic ash is weathered, so was probably around when the Earth was young and pockmarked with active volcanoes.

Two things make montmorillonite special. The first is that molecules like nucleotides attach to its surface, forming a thin film. This is called 'adsorption': it's like absorption, except the nucleotides don't get sucked inside the clay. And the second is that montmorillonite is a natural catalyst, speeding up chemical reactions among the molecules that are stuck to it. It contains metals, like magnesium, which are often found in enzymes. Much of this had been suggested forty years before by Bernal in a 1947 lecture,† but it was Ferris who followed through on it.

From 1986 onwards, Ferris produced a stream of experiments demonstrating that nucleotides on montmorillonite are more likely to link up.[23] In 1996, with Orgel, he showed that nucleotides would form chains fifty-five long if they were on montmorillonite and extra nucleotides were plentiful.[24] That was significant, because RNAs had to be long to carry significant information or to form ribozymes. Montmorillonite clearly helps – but Ferris never made a self-replicating RNA on it.

A second tack was to suppose that the first genetic material was not exactly RNA, but some slightly modified version that was more amenable to self-replication. Orgel had been toying with this since the 1970s. In 1985 he and Schwartz made chains of modified

* Told you so. (See chapter 5.)

† This was the same lecture in which Bernal argued that the primordial atmosphere was probably not reducing, as we saw in chapter 6.

nucleotides, in which the phosphate had been stripped off and replaced with a similar molecule.* These modded nucleotides could link up, if they first aligned themselves on an RNA-like template.[25] Two years later, Orgel issued a rallying call for alternative nucleic acids. With Schwartz, Gerald Joyce and Miller, he argued that 'there are several reasons for doubting that RNA itself was the first genetic material'.[26] It was an extended 'Yeah, but . . .' directed at the newfound popularity of the RNA World.

Another biochemist thinking along similar lines was Günter von Kiedrowski, who studied under Orgel in the 1980s. He has created remarkable self-replicating systems. For example, in 1993 his team devised a system based on three modified nucleic acids dubbed A, B and C. These could link up to form AB, AC, BC, BB and ABC.[27] When the team added some short nucleic acids, a dizzyingly elaborate set of reactions unfolded. This was an autocatalytic set as Kauffman had envisioned it, with multiple nucleic acids each triggering the formation of another, like a flowing dance routine choreographed by Busby Berkeley.

Many of the molecules Orgel and von Kiedrowski used were similar to the familiar components of RNA and DNA.[28] But others ventured further afield.

In 1991, Peter Nielsen of the University of Copenhagen invented a new nucleic acid that has never been found in nature.[29] He kept the bases – the 'letters' A, T, C and G – but threw out the sugars and phosphates that form the backbone of DNA and RNA. He replaced them with amides, which are similar to amino

* A nucleotide that has had its phosphate removed is called a nucleoside. The similarity of the names is presumably a trap for time-poor science journalists.

acids. Nielsen initially called his creation 'polyamide nucleic acid', but it has since become known as 'peptide nucleic acid'. Either way, it's normally shortened to PNA.

Despite its radically different structure, PNA behaves in familiar ways. It can form a double helix like DNA.[30] It is also possible to make hybrid molecules where one strand is PNA and the other DNA. Furthermore, the amides in the PNA chains are simpler molecules than nucleotides – arguably making them more likely to form naturally.

Stanley Miller was a fan. In 1997 he wrote that the discovery of PNA had been 'a complete surprise' and that it 'looks promising', whereas RNA was 'an unlikely candidate'.[31] Three years later, in one of his last published papers, he performed a variation on his original experiment in which he obtained the amides that make up the PNA backbone. He argued that PNA might have formed on the early Earth – although this experiment involved a methane-rich reducing atmosphere, which as we saw in chapter 6 probably never existed.[32]

PNA is not the only newly created nucleic acid proposed as an alternative to RNA. The other prominent one is threose nucleic acid (TNA), first made by Swiss chemist Albert Eschenmoser and his team in 2000.[33] It is less radical than PNA: instead of chucking both the ribose sugar and phosphate, the ribose is replaced with a substitute sugar called threose. Like PNA, TNA can form a double helix. It can also base-pair with RNA. More impressively, TNA can fold up into complex three-dimensional shapes.[34] This implies it might be able to form enzymes, like RNA.*

* Presumably we would have to call these 'threozymes'.

Thanks to these discoveries, by the turn of the millennium the idea that some other polymer preceded RNA had gathered a head of steam. But there were, and are, plenty of counter-arguments. The first is simply that none of the alternative molecules have been found in any living organism. While this could mean that RNA and DNA completely took over early in the history of life, it could also mean that PNA and TNA never played any role. Furthermore, just because their molecular structures *look* simpler to us doesn't necessarily mean they were more likely to form.

As a result, many kept faith with the RNA World despite the difficulties. Then, in 2000, the RNA World was gifted perhaps its most dramatic piece of evidence.

The new discovery centred around the ribosomes, the massive molecular machines that assemble proteins by following the instructions on messenger RNA. Ribosomes are essential for all known organisms and have existed since the time of the Last Universal Common Ancestor. It had been clear for decades that ribosomes were made of protein and RNA, tangled together – and that each ribosome contained two subunits of different sizes. But in 2000 the precise atomic structure of the ribosome was unveiled.

This was the culmination of two decades of effort. It began in 1980 when Ada Yonath of the Weizmann Institute of Science in Israel managed to crystallise part of the ribosome.[35] This meant the ribosome could be studied using X-ray crystallography, the method Rosalind Franklin used to help unravel the structure of DNA. By 1984, Yonath had her first X-ray images.[36]

Over the next fifteen years, researchers including Thomas Steitz and Venkatraman Ramakrishnan followed Yonath's lead.

The X-ray images became more precise. Finally, in 2000, Steitz's group published a detailed structure of the large subunit, showing that the active core of the ribosome was made up of RNA.[37] Or, as Thomas Cech succinctly put it, 'the ribosome is a ribozyme'.[38] Weeks later, Yonath and Ramakrishnan's groups revealed similarly detailed models of the small subunit, completing the picture.[39] In 2009, all three shared the Nobel Prize in Chemistry.[40]

It would be hard to overestimate the significance of this discovery for the RNA World hypothesis. It was an obvious smoking gun. The ribosome is one of the most central components of living organisms, and its active core is made of RNA. This implies the first ribosome could have been made of nothing but RNA. These crude ribosomes may have simply been small RNAs that bonded to amino acids. When two such RNAs paired up, the amino acids could have been pushed together and linked up.

Since 2000, the RNA World hypothesis has enjoyed several additional coups, which have bolstered it further. Another Orgel student, Gerald Joyce, has made startling progress on the problem of self-replicating RNA. Taking a cue from von Kiedrowski, Kauffman and their networks of molecules, Joyce has abandoned the idea that a single RNA might copy itself. Instead, he has developed pairs of RNAs, each of which makes the other – in other words, extremely simple autocatalytic sets.

Over the 2000s, Joyce, with colleagues Dong-Eun Kim and Tracey Lincoln, tinkered with a ribozyme that ordinarily links two short RNAs together to form a longer strand.[41] In their version, the ribozyme joined two short pieces of RNA together to form a different ribozyme. This second ribozyme in turn would stitch together two RNAs to make a copy of the original ribozyme. Given enough raw material, in the form of short RNAs of the correct sequences, this could continue indefinitely. One might ask where the short RNAs came from, but as a proof of principle it was a beautiful result.

Similar studies have emerged in recent years. They show that RNA really can self-replicate. It just does not do it in the straightforward way envisioned by Orgel and Gilbert, in which a single ribozyme copies itself. Instead, there have to be several different ribozymes that work together – and such co-operative groups can form.

This is good news for the RNA World hypothesis. Along with the nature of the ribosome, and the ability of montmorillonite to encourage the formation of RNA from nucleotides, the hypothesis has passed a lot of critical tests.

Nevertheless, biochemist Harold Bernhardt spoke for many weary researchers when he wrote that the RNA World hypothesis was 'the worst theory of the early evolution of life (except for all the others)'.[42] In particular, many doubt that nucleotides could have formed without the aid of enzymes, and no nucleotides means no RNA. This explains the persistence of those researchers who favour alternative nucleic acids like PNA.

To some extent, how people view the RNA World hypothesis depends on which version they are talking about. The 'hard' version, as set out by Gilbert, is that the first life was based entirely on RNA, with no proteins or other macromolecules involved. This looks decidedly shaky. Nobody has demonstrated that ribozymes can run an entire metabolism. However, many researchers do accept a 'soft' version. In this scenario, RNA is still the first genetic material, preceding DNA, but it does not have to do everything else on its own like some ridiculous superhuman from an Ayn Rand book. Other macromolecules like proteins were also present and played key roles.

Still, even this iteration of the RNA World raises difficult questions. In particular, if the RNAs were floating in a primordial soup, it is hard to understand how they could have stayed together instead of just drifting apart. The same is true for RNAs clinging to montmorillonite clay: a good downpour would wash them away. This led some researchers to argue that the whole thing was a non-starter unless there was something that could hold the key molecules together.

Chapter 9

The Blobs

So far we've seen two alternatives to Oparin's primordial soup hypothesis: Fox's notion that proteins came first and the RNA World. Both have several things going for them, but both are also manifestly incomplete.

The third school of thought is that life began with a container or compartment of some sort: something that could hold all the other bits together once they arose. This is called the 'compartmentalisation-first' model. It generally means a crude version of the outer layers or membranes that today hold in the contents of cells. These objects are often called 'protocells', but that can be a little confusing because it implies they have all the working parts. What is being talked about is something like a bubble or droplet. So long as it can hold other chemicals inside itself, keeping the living system intact, it will do.

The other key advantage of protocells is that they allow evolution by natural selection to take place, because each protocell is a distinct individual that can compete against its neighbours. If instead there is a mess of RNA, proteins and the like all jostling around in a primordial soup, there are no individual organisms. One molecule can compete against another for survival, but that is not guaranteed to lead to greater complexity.

The history of the compartmentalisation-first idea is tangled. In part this reflects the course of research into the outer layers of living cells. Although the outer layer was the first part to be discovered – it was all Robert Hooke could see when he looked at cork cells under a microscope – it would not be properly understood until the 1970s.[1]

The first difficulty was that not all cells have the same outer layer. Plant cells like the ones Hooke looked at have thick and rigid 'walls', visible under a crude microscope, but these are unusual, and in evolutionary terms they were a late invention. Instead, most cells have thin and flexible outer layers, known as membranes or sometimes plasma membranes. They were so tricky to see under microscopes that for much of the 1800s there was a spirited argument about whether cells needed membranes to hold themselves together, or if membranes even existed. This was not settled until the early 1900s, by which time it was clear that the membranes prevented some chemicals moving in and out of cells. A membrane was far from an absolute barrier, but it was more of an obstruction than the jelly inside the cell.

The challenge was to figure out what the membrane was made of and how it was structured. The first person who properly tackled this was British biologist Charles Ernest Overton. In the 1890s, after a lengthy series of experiments, he correctly concluded that the cell membrane was made of chemicals called lipids. This is a huge group of molecules and includes some household names: notably fats, oils and cholesterol.* The thing that unites the lipids is that they all have long chains of carbon atoms. They often look a bit like a tadpole, the 'head' being some complicated molecule and the 'tail' being the carbon chain dangling from it.

Lipids have a peculiar love–hate relationship with water. Most people have heard the saying that oil and water don't mix, and this is easily confirmed by adding a few drops of cooking oil to a jug of water: the oil will form either a thin film on the water's surface, or spherical droplets underneath.

This happens because of the way electrons are spread around on the molecules. On a long chain of carbon atoms, the electrons are evenly spread so no part of the molecule has a strong electrical charge. But on a water molecule, things are more lopsided. The electrons cluster towards the oxygen, which therefore has a negative charge, while the two hydrogens each have a positive charge. Chemists say water is a 'polar' molecule, while the carbon chain of a lipid is 'non-polar'. When the two meet, they repel each other. The non-polar lipid chain would rather sit next to another non-polar molecule, and the polar water molecule prefers to cuddle up to another polar molecule. Molecules that 'dislike' water are said to be hydrophobic; those that happily mix with it are hydrophilic.

* The terminology of lipids can be a nightmare, because different kinds of scientist use the words differently. Technically, 'lipid' is the umbrella term and fats are a subgroup, but nutrition scientists will sometimes say 'fats' to mean 'lipids'. Also, fats and oils are supposed to behave differently: the former are solid at room temperature while the latter are liquid. As far as possible, I'll just use 'lipid' and be done with it.

This explains why Overton thought cell membranes were made of lipids. They naturally form bubble-like objects when mixed with water, so they seemed a good fit for creating a spherical barrier around a cell.

However, there was a further wrinkle. Overton thought – again correctly – that cell membranes were probably made of a specific kind of lipid called a phospholipid. In these molecules, the head of the molecule is a phosphate group (just like the ones in DNA, RNA and ATP). Phosphate is a polar molecule, so phospholipids have a polar head and a non-polar tail. That means their heads are hydrophilic and like being next to water, but their tails are hydrophobic and avoid it. If you put phospholipids in water, they instantly form neat spheres. The phospholipids arrange themselves with the water-loving heads pointing outwards and the water-hating tails nestled safely inside. They were clearly the perfect molecule for a cell membrane.

The next big step was taken by two Dutch scientists in 1925. Evert Gorter was a paediatrician who tackled infant mortality, but he also studied biochemistry alongside his colleague François Grendel. They removed all the phospholipids from cells and spread them out into layers a single molecule thick. The phospholipids from each cell had a surface area twice that of the cell, so Gorter and Grendel concluded that the cell membrane must contain two layers of phospholipids sandwiched together.[2] This was correct and soon became broadly accepted.*

* It emerged later that Gorter and Grendel only got the right answer by accident. Their experiment had two sources of error, which fortunately cancelled out to give the right answer.

By the 1930s it was clear that membranes were made of a double layer of phospholipids, but there was a further complication: membranes also contained proteins. This led British biologists James Danielli and Hugh Davson to suggest that the phospholipid bilayer was sandwiched between layers of proteins, an idea that held sway for over thirty years.[3] Not until 1972 was the correct solution put forward by two Americans, Seymour Singer and Garth Nicolson. They argued that the proteins are not smeared onto the outside of the phospholipids like butter on bread, but instead are embedded in the phospholipid layers like chocolate chips in cookie dough.[4] Some proteins extend all the way through the membrane, their ends jutting out on either side. They can act as channels or pumps, funnelling chemicals from one side of the membrane to the other.

This picture of the cell membrane as a double layer of phospholipids, with proteins poking through it at intervals, is still generally accepted over forty years after it was proposed. Only once this was established was it possible to develop a fact-based hypothesis of how the first membranes formed.

Not that this stopped people trying. We saw in chapter 2 that Oparin believed the first cells were droplets called coacervates, which did not have a defined outer membrane. This raised two difficulties: were the coacervates resilient enough for life to begin within them, and how were they replaced by true membranes? Later, Fox developed his proteinoid microspheres, which he suggested were the progenitors of cells. Again, it's unclear how they were replaced by phospholipids.

One of the most idiosyncratic attempts at making the soft exterior of a cell was performed by an Indian scientist named Krishna Bahadur at the University of Allahabad.[5] He often worked with his wife, S. Ranganayaki, whose first name does not seem to be recorded. In 1954, a year after Miller's classic study, Bahadur had described an alternative way to make amino acids from simple chemicals.[6] This earned him an invitation to the 1957 Moscow conference, although he could not attend.

Over the next few years, Bahadur's team created cell-like particles that he called 'Jeewanu', from the Sanskrit words *jeewa* ('life') and *anu* ('the smallest part of something'). Bahadur made the Jeewanu by mixing various chemicals, then shaking them and illuminating them with sunlight. There was no single recipe: sometimes Bahadur used proteinoids similar to Fox's, but he also used less lifelike chemicals such as copper oxide. Each Jeewanu blob was surrounded by a layer that superficially resembled a membrane.

Bahadur first described the Jeewanu in 1963 in a little-known Indian journal.[7] He published four detailed papers in an equally obscure German journal the following year.[8] Biologists who noticed were intrigued but cautious. Bahadur laced his papers with quotes from ancient Hindu scriptures called the *Vedas*, and explicitly linked the Jeewanu to Hindu thought on the nature of life: in particular the idea that there was no stark boundary between life and non-life. He viewed his little particles as alive, claiming that they had metabolic activity and could reproduce. Whereas other scientists have been reluctant to describe simple lifelike systems as truly alive, Bahadur seemingly had no hesitation. It is understandable that people were sceptical.

Nevertheless, NASA took a look: Cyril Ponnamperuma and his colleague Linda Caren were assigned to review Bahadur's findings.[9] They criticised the 'confusing' and 'inadequate' descriptions of the experiments, which made them difficult for anyone to repeat. They also stated that 'insufficient evidence has been advanced to prove that jeewanu are alive'. In other words, Bahadur was overstating his results. Ponnamperuma and Caren did not dismiss him outright, concluding that 'the nature and properties of the jeewanu remain to be clarified', but their critique was influential and his ideas never gained currency.[10] Since Bahadur's death, his former student Vinod Kumar Gupta has continued studying the Jeewanu, producing papers in which he claims, for instance, to have detected 'phospholipid-like substances' in the outer surface – reinforcing the similarities between Jeewanu and living cells.[11] Nevertheless, the Jeewanu remain a curiosity rather than a milestone.

The first researcher to demonstrate a protocell with a membrane similar to those of living cells was American biologist David Deamer. Born in 1939, Deamer began studying membranes in the 1960s. He has had a remarkably successful career. In 1989 he was driving alone through Oregon when it suddenly occurred to him that it might be possible to read the sequence of bases on a DNA molecule by using an electrical field to force it through a protein channel embedded in a membrane. The four different bases would produce characteristic electrical fields as they passed through. He hastily pulled up by the side of the road and jotted the idea down before driving on. Deamer and colleagues later developed this into a technology called nanopore sequencing. Portable versions have since been used on the International Space Station.

Deamer began thinking about the origin of membranes in 1975, a few years after the correct model of membranes was published. That spring, he was on sabbatical at the Babraham

Institute near Cambridge in the UK, with fellow lipid specialist Alec Bangham.[12] A few years earlier, Bangham had given a lecture titled 'Membranes Came First'. 'Alec and I were driving down to London in his little Mini Cooper and had stopped for lunch,' Deamer recalls. 'We realised nobody had actually tried to think about how membranes came to be on the early Earth.' Somehow the first life must have acquired membranes, but how?

Deamer's graduate student William Hargreaves seized on the question. In 1977, Hargreaves, Deamer and a colleague published an experiment suggesting that phospholipids could have formed from simpler chemicals on the young Earth.[13] They envisioned tide pools on sandy shores, with chemicals dissolved in the water. They focused on glycerol: a simple alcohol that can be used to make lipids. Also known as glycerine, it is used as a sweetener in foods. By mixing it with the reactive compound cyanamide and water, and heating the mixture to 65°C, they made several lipids including phospholipids.

They also confirmed that these phospholipids could perform the key trick phospholipids were known for: assembling into cell-like structures. They dissolved the newly made phospholipid gunk in salt water and gave it a shake. The phospholipids gathered into tiny spherical blobs that looked like cells. Closer inspection revealed that each blob had an outer membrane composed of a double layer of phospholipids, just like that of a living cell. Such structures are called vesicles and all cells have them: they are used to store useful things like food molecules.

Deamer's team had evidence that simple cell-like structures, made of the same sort of chemicals that modern cells use for membranes, might have formed naturally billions of years ago. As with all such experiments, one can question whether the chemical reactions they used would really have happened. However, in this case it seems reasonable. Cyanamide and glycerol are simple, common molecules, so it would not be surprising if they were present, and there are plenty of places on Earth where water is heated to 65°C.

Similar experiments with other kinds of lipid followed over the next few years.[14] By 1980 Deamer could write: 'Membrane structures resembling contemporary membranes are readily produced from lipid compounds likely to have been available on the prebiotic Earth.'[15]

However, a vesicle is not a cell. Making the outer membrane and persuading it to form a nice spherical shape was one thing – but so far the vesicles Deamer and his colleagues had made could not do anything much. The next step was to see whether they could store biological molecules like DNA inside them, something Deamer tackled with Gail Barchfeld. Once again, they envisioned pools of water on land. Such ponds often dry out in the hot sun, then fill up again when it rains – and so on. Deamer and Barchfeld mixed phospholipid vesicles with DNA and put them through these dehydration–rehydration cycles. When the vesicles were dried out, they changed shape drastically: the phospholipids rearranged themselves into flat layers like a stack of pancakes, trapping DNA between them. Later, when the water returned, the vesicles reformed – but now each contained DNA. These protocells were still not alive, but Deamer and Barchfeld had brought them a small step closer to it.[16]

By now Deamer was far from alone in pursuing a compartmentalisation-first view of the origin of life. Over the course of the 1980s other researchers became involved.[17] One was biophysicist Harold Morowitz, who had been thinking about the origin of life since the 1960s and whose ideas will pop up in later chapters. By the late 1980s he was convinced Deamer was on the right track and that simple cells – probably vesicles like those Deamer had made – must have emerged first.

In 1988 Deamer and Morowitz, with biochemist Bettina Heinz (who has since become a successful artist)[18] argued that such vesicles constituted 'minimum protocells': the simplest possible form of cell.[19] To become more lifelike, the vesicles needed a source of energy. The team speculated that they could have had coloured pigment molecules embedded in their membranes. When pigments are hit by sunlight, they release electrons, which can then trigger chemical reactions – for instance, forming new lipids to further build up the membranes. Deamer went on to explore specific groups of pigments.[20]

Morowitz developed his own ideas further in his 1992 book *Beginnings of Cellular Life*.[21] His thesis was that even a simple vesicle possesses many of the properties of a cell. The outer membrane is a surface on which chemical reactions can be catalysed. Chemicals can be concentrated inside the vesicle, or kept out if they are undesirable. Morowitz was convinced that vesicles formed early, with metabolic reactions coming soon after and genes last.

The key element of Morowitz's hypothesis was that the vesicles acted as a foundation, driving the formation of the other components of life. It was not simply 'and then proteins formed'. Rather, the vesicles would pick up additional molecules that were sensitive to light, embedding them in their membranes. These extra molecules would then become electrically excited when the sun shone on them, creating patterns of electrical charge that could trigger new chemical reactions. It was this that kick-started metabolic reactions by converting carbon dioxide into more elaborate chemicals, ultimately making things like amino acids.

Today, Morowitz's idea that the first organisms got their energy from light, rather than from natural chemical flows, looks a little unlikely. That is because, thanks to molecular genetics and the attempts to reconstruct the Last Universal Common Ancestor, we now have a better understanding of what early microbial life was like. It seems the oldest microbes are those that feed off natural chemical reactions, and the ability to harness sunlight evolved later. This was not known when Morowitz was writing.

The other researcher who advocated protocells in the 1980s and 1990s was Pier Luigi Luisi. An Italian biochemist who spent most of his career in Switzerland, Luisi is something of a polymath who has written on the philosophy of science and whether nature can be said to have a purpose.[22] He co-authored a book on the nature of reality in which he variously consulted the Dalai Lama and the actor Richard Gere.[23]

In the early 1980s, Luisi's team studied micelles: spherical blobs of lipids that look at first sight like vesicles. However, micelles only have a single layer of lipids on the outside, rather than two. In 1989, Luisi and a colleague suggested that micelles could be self-sustaining, if they could produce new lipids inside themselves.[24] Such a process might even allow the vesicles to self-replicate: to reproduce.

From 1990 to 1992, Luisi's team demonstrated ways to do this. Their first attempt used micelles made of a fatty acid eight carbon atoms long, called octanoic acid, plus a near-identical chemical called octanol. Inside the micelles was a third, more complex molecule called octyl octanoate. This broke down when it reacted with water, forming both octanoic acid and octanol. When these molecules escaped into the water, they spontaneously assembled into a new micelle, so instead of one micelle there were now two.[25] This was a simple form of reproduction. It was artificial, as it used chemicals unlike those in real cell membranes and did not mimic the process of cell division, but it was striking nonetheless.[26]

The team soon found that many lipids could make micelles and replicate in a similar way. But, more importantly, they found that it was not necessary to start with pre-existing micelles. Instead, they could start with ethyl octanoate mixed with alkaline water. The ethyl octanoate reacted with the water to make sodium octanoate, which formed micelles. The first micelles swiftly attached themselves to the remaining ethyl octanoate molecules, speeding up their destruction and thus creating more micelles.[27] This was a process by which a population of micelles could assemble from scratch.[28]

In the second half of the 1990s, Luisi turned his attention to the processes that could be hosted inside micelles and vesicles. Could these crude versions of cells host a working enzyme, or the copying of a nucleic acid like RNA? If they could, it was a further hint that they could have been the progenitors of cells.

As it turned out, many biological processes work perfectly well inside these protocells. For example, vesicles can enclose the enzymes that make more of their constituent lipids.[29] Similarly, Deamer's team demonstrated that RNA could be made inside a vesicle, given nucleotides and the necessary enzyme.[30] Most dramatically, in 1999 Luisi's group showed that vesicles could host ribosomes, despite ribosomes being gargantuan by molecular standards.[31] Furthermore, the ribosomes could successfully build proteins if they had a supply of amino acids.

These studies should again be seen as a proof of principle. It seems unlikely that the first cells would have evolved a big, modern ribosome while retaining a simple outer membrane: it would be like equipping a 1980s PC with a 2020 supercomputer processor. But the essential point, that the processes of life are robust enough to work within a simple protocell, stands. 'These experiments can be seen as a first approach towards a minimal cell,' Luisi wrote.[32]

To many researchers, it now seemed quite reasonable that vesicles or micelles could have formed spontaneously on the primordial Earth, and even begun to self-replicate. These first protocells may not have been made of the exact same lipids as modern cells, but Luisi and Deamer had shown that that need not have been a problem for them. Furthermore, once some additional cellular machinery had evolved, the original crude lipids could have been gradually phased out by evolution, replaced by the more suitable phospholipids.

Still, there was a glaring problem. How did these protocells become associated with the other components of life? Suppose for a moment that an RNA World got started somewhere, perhaps in a geothermal pond. How and why would it have joined up with the lipid vesicles – which might have formed somewhere else? The problem was that an RNA World might originate in one locale, a vesicle world in another, and some sort of metabolic cycle based on crude enzymes in yet another location. Without RNA, proteins and the other components of life, the vesicles were stuck. They could progress a little way towards becoming alive, but not all the way.

The Israeli researcher Doron Lancet has argued that this was not the dead end it seems to be. Most of Lancet's career has been spent on genetics and the sense of smell, but in later years he has developed an interest in the origin of life. Since 2001 he has argued that life began in what he calls the Lipid World.[33] Along the lines of Gilbert's suggestion that RNA could perform all the key functions of life, Lancet argues that lipids could stand alone.

The basic idea, which Lancet developed alongside several colleagues including Deamer, is that lipid vesicles may be able to pass on a kind of information to their descendants, even if they do not have any conventional genetic molecules like DNA. In other words, vesicles made of nothing but a mix of lipids could compete with each other and evolve in a rudimentary way.

How might this work? The first point is that there are many different lipids, and the chemical evidence suggests many of them were present on the young Earth. This implies that the first vesicles would not have been made purely of one kind of lipid, as the membranes of modern cells are, but rather a mixture. Such vesicles have been made in the lab and often work fine. In this scenario, it's likely no two vesicles had exactly the same mix of lipids.

The second critical factor is that vesicles can have a catalytic effect: their presence can drive chemical reactions forward. This can sometimes happen simply because chemicals get trapped inside a vesicle, crushing them together and making them more likely to react. Just as catalytic RNA molecules have been dubbed 'ribozymes', Lancet has proposed calling catalytic vesicles 'lipozymes'.

From these premises, Lancet has developed a scenario for a wholly lipid-based origin of cells. Any vesicles that catalysed the formation of their own constituents would reproduce themselves and become more common. The mix of lipids in these vesicles would become more complex as they acquired more catalytic properties, until eventually they started catalysing the formation of other biological molecules like proteins. At that point, these purely lipid-based protocells would start to look and behave more like modern cells.

Lancet's team has produced mathematical models of these hypothetical lipid communities, extrapolating how they would behave. These simulations indicate that the vesicles could indeed evolve to become more complex. However, this has not been verified with practical experiments. Lancet says there are 'insurmountable hurdles' to doing so, because of the complexity of the chemistry.[34] The complexity is certainly challenging, but the lack of even a crude attempt probably accounts for the model's relative obscurity.

It is clear that the Lipid World represents a kind of endpoint. Like Fox's proteinoid microspheres and the 'hard' RNA World, it is an attempt to create all the key functions of life using only one class of molecules. For the model to work, those molecules have to do many things they are not usually called on to do. However, for most origins researchers the idea of trying to make life using only one kind of biological molecule is now passé and unrealistic. Nevertheless, the Lipid World remains an important idea, if only because it explicitly illustrates the limits of what can be achieved with a single class of molecule. That is, it is probably wrong, but its failure points towards a better solution.

Meanwhile, Deamer and Luisi's studies of lipids and protocells are critical. They reveal that it is relatively easy to make the superstructure of a cell from a bare-bones set of chemicals. This tells us that protocells made of lipids may well have played a role in life's origin.

But first, we must tackle the fourth and last of the competing theories that evolved in the late twentieth century: an approach that was literally energetic, and whose proponents would force everyone else in the field to reconsider their most basic assumptions.

Chapter 10

The Need for Power

Leonard Troland's life story is so extraordinary that it is hard to believe he isn't better known. After graduating from Harvard in 1912 and MIT in 1915, he studied both psychology and fundamental physics. These seemingly disparate interests were linked by Troland's lifelong fascination with light. He sought to understand both our sense of sight and the nature of light. He may even have been the first to use the term 'photon' for the smallest particle of light, in 1915: eleven years before it was 'officially' proposed.[1]

The following year Troland conducted one of the first scientific studies of telepathy, finding no evidence that it exists – a conclusion that still holds firm. In the 1920s he somehow found time to become chief engineer at the Technicolor Motion Picture Corporation of California. According to one profile, he 'not only developed and improved the old two-color process of color photography, but he invented and perfected the modern multicolor process in all its details'.[2]

However, this frantic work took its toll. By early 1932 Troland had suffered a nervous collapse from overwork, causing dizzy spells and fainting.[3] On 27 May he climbed Mount Wilson in California with a friend. On the summit he posed for a photo, but as his friend looked through the viewfinder, Troland vanished. He had fallen hundreds of feet into a rocky canyon, to his death. He was only forty-three.

It is a mark of Troland's fertile mind that his speculations about the origin of life were published in 1914, before he had even completed his PhD – in psychology.[4] He seems to have been the first scientist to outline a scenario for life's origin that focused on metabolism: life's ability to obtain energy from its surroundings and use that energy to maintain itself. The idea that metabolism came first is the fourth and last of the major competing hypotheses. There are several significant variants, covered in both this and the following chapter.

Troland argued that the most essential feature of living organisms is their ability to keep themselves pretty much the same. 'Regulation seems to be the most striking active characteristic of living beings,' he wrote. To him, this was the central problem that any theory of life's beginning had to solve. How did the first living thing stabilise itself? His answer was that certain chemical reactions had to happen while others had to be prevented. This suggested enzymes were essential for the emergence of life.

Troland imagined that, early in Earth's history, an enzyme appeared by chance in the ocean. Alongside it there were chemicals that reacted slowly to produce 'an oily liquid' – a lipid, in other words. If the enzyme accelerated this reaction, there would only be one outcome: 'Why, obviously the particle of enzyme will become enveloped in the oily material resulting from the reaction.' The result would be a 'little oil drop' with an enzyme at its core, which Troland called 'the first and simplest life-substance'.

Troland went on to outline how an enzyme in an oil drop could have evolved into more complex living cells. He argued that the first enzyme would ultimately lead to the formation of more, which would drive new chemical reactions. These would either help or hinder the oil droplet. Only droplets with helpful enzymes would survive, so a simple form of evolution would take place and the 'cells' would 'become more and more complex'.

For Troland, life was almost a secondary phenomenon, something thrown up as a side effect of the existence of enzymes. 'Life,' as he put it, 'is a corollary of enzyme activity.'

Today, the specifics of Troland's idea seem unlikely. We saw in chapter 6 that we cannot rely on pure chance for the formation of anything so complicated as an enzyme: there has to be a repeatable process. Of course, this is not Troland's fault: he was writing before the structure of protein enzymes was understood, and many other researchers invoked the laws of chance in similar ways. In any case, the value of his idea is conceptual: it focuses on life's ability to build and maintain its structure, despite the chaos of its environment.

Troland's emphasis on life's ability to regulate itself can be put on a more rigorous footing by considering it in terms of thermodynamics. This branch of physics was created to deal with the nature and behaviour of heat, and ended up yielding some of the most profound revelations in all of science.

The bit that relates to the origin of life is the Second Law of Thermodynamics, which essentially says that the universe inevitably becomes messier over time. This is because processes that take orderly things and make them disorderly are overwhelmingly likely to happen, whereas order does not spontaneously emerge from disorder. For example, an intact coffee cup is an ordered phenomenon. It is easy to turn it into a smashed coffee cup, which is disordered, but it is hard to take that smashed coffee cup and turn it back into an intact one. Crucially, we never see this second process happen spontaneously. Broken cups do not pick themselves up off the floor and reassemble.

This intuitive truth can be put into more scientific terms by measuring the 'entropy' of a system. Entropy is not a physical thing like an atom or the force of gravity, but rather a mathematical abstraction that tells us how disordered something is. An intact coffee cup has low entropy and a broken one has high entropy. The Second Law tells us that, overall, entropy always increases.

This becomes clear if you consider a slightly more artificial setup. Imagine two chambers, one of which holds particles of a blue gas and the other holds particles of a red gas. This is an ordered system with low entropy. Now open a door between the two chambers. We all know what will happen: the two gases will mix, so the entropy will increase. It is theoretically possible that at some point, by sheer chance, the two gases will separate again, decreasing the entropy – but you will have to wait longer than the age of the universe to have any reasonable prospect of seeing it. The same applies to the coffee cup: it is not impossible for a broken coffee cup to reassemble itself, just vanishingly unlikely.

At this point, it may seem that there is an obvious exception. Sure, you might smash a coffee cup and increase its entropy, but you can patiently glue it back together and bring the entropy back down. However, the Second Law is ahead of you. All that work you do to fix the cup will release heat, which causes atoms in the surrounding air to move around in a disordered way. So you may well reduce the entropy in the coffee cup, but you will inevitably cause entropy to rise somewhere else – and that rise in entropy will always be bigger than the fall in entropy in the cup. At the scale of the universe, the disorder always wins. You may create a little pocket of tidiness, but outside it things will inevitably get more disorderly.

Whatever hope you have for the ultimate fate of the universe, surrender it now. The Second Law tells us that the universe will inevitably end up in a disordered mess in which it's no longer possible for anything interesting to exist or happen.*

There is no getting out of this, because the Second Law is absolute. Most scientific discoveries are in some sense always provisional, because they might be overturned by some future discovery, but this is not true of the Second Law. That is because it follows from the fundamental mathematics of probability, so it can be definitively proved true in a way most ideas can't. This is not like saying, 'All swans are white – oh, hang on, there's a black one.' The only way the Second Law can be wrong is if basic maths, which we all experience as true, is somehow wrong.

An easy way to see this is to throw ten dice at once. They might come up in an orderly way: all sixes, say. But there is only one way to roll all sixes, and vastly more ways to get a messy combination like 5364414235. The sheer weight of numbers means you will almost always get a disorderly result.

* Unless something else dreadful happens first. Physicists have come up with a number of exciting scenarios for the end of the cosmos, with names like Big Crunch and Big Rip. They make *The Road* look like *Care Bears*.

The physicist Arthur Eddington expressed this clearly. He wrote that a scientist's pet theory might survive contradictory experiments, because 'these experimentalists do bungle things sometimes', or a clash with long-established ideas. 'But if your theory is found to be against the second law of thermodynamics I can give you no hope; there is nothing for it but to collapse in deepest humiliation.'[5]

Scientists can get quite melancholy about the consequences of the Second Law. It's perhaps not a coincidence that Ludwig Boltzmann, who first spelled out the law's true meaning, hanged himself. The chemist Peter Atkins opened his book *The Second Law* with this cheerful announcement:[6] 'We are the children of chaos, and the deep root of change is decay. At root, there is only corruption, and the unstemmable tide of chaos. Gone is purpose; all that is left is direction.' Atkins would have been an asset to any early 2000s emo band.

What does this tale of ever-increasing disorder have to do with life? Once again, physicist Erwin Schrödinger spelled it out in 1944 in *What Is Life?*: 'Life,' he wrote, 'seems to be orderly and lawful behaviour of matter, not based exclusively on its tendency to go over from order to disorder, but based partly on existing order that is kept up.' In other words, life is very good at maintaining itself, at staving off disorder. For instance, your hair probably doesn't spontaneously change colour every few hours.* But it is more complicated than simply staying the same:

* Unless your name is Nymphadora Tonks and you are a character in *Harry Potter.*

rocks do that. In contrast, life is energetic, constantly in motion. Even a tree, while it may look sedentary, is a hive of activity at the cellular scale. Life keeps on doing things 'for a much longer period than we would expect of an inanimate piece of matter', as Schrödinger put it.

At first glance, it might seem as if life is breaking the Second Law, but it isn't. Life has to work to keep going, so it puts out both heat and waste chemicals – think of how much poo you've produced in your life. The entropy in your body might stay low, but you are exporting entropy to the rest of the universe.

'How does the living organism avoid decay?' asked Schrödinger. 'The obvious answer is: By eating, drinking, breathing and (in the case of plants) assimilating. The technical term is metabolism.' In other words, all living things need a supply of energy so that they can do the work of maintaining themselves. Or, to put it another way, by running a metabolism an organism 'succeeds in freeing itself from all the entropy it cannot help producing while alive'. This is what Troland was getting at when he described life's ability to 'regulate' and 'maintain' itself.

Today, many researchers are convinced that some form of metabolism must have appeared first, before any other aspect of life. The argument is simple: without metabolism, without a way to harness a source of energy, life can't maintain itself. It's all very well having RNAs that can replicate themselves, but it won't be much good if they fall to pieces first.

The energy in living organisms isn't a bolt of lightning sparking through their cells. Instead, all living cells use a molecule called adenosine triphosphate (ATP) to store energy.* Every organism, from the simplest bacteria to humans, uses ATP: it is a universal energy 'currency'. When you move a muscle, or a brain cell fires, or a bacterium divides into two, energy from ATP makes it happen.

ATP works a bit like a rechargeable battery. First, an organism gets energy from some source or other. It uses that energy to bolt phosphates onto adenosine until there is a chain of three, making ATP. At this point, the energy is safely stored in the chemical bonds linking the phosphates. Then, when the organism needs energy to do something, it strips phosphates from the ATP, releasing the energy.

This process of constantly making and unmaking ATP is fundamental to life on Earth. It is as universal as proteins and nucleic acids. Your cells are doing it right now, and so is every bacterium on your skin.

So there are good reasons why many people believe that life must have begun with a way to obtain and use energy. The idea has a long pedigree. In 1963, a decade after Miller's experiment, biochemist Robert Eakin suggested that something similar to ATP must have been present in the first living organisms.[7]

However, it took a long time for someone to come up with a plausible primordial metabolism. Troland had imagined a single enzyme, but that is unrealistic. In modern organisms, even simple bacteria, metabolism involves long series of chemical reactions. These often work in cycles: a starter chemical gets changed into a different one, which gets changed into a third, and so on until

* ATP previously came up in chapter 3, when Cyril Ponnamperuma and Carl Sagan tried to show that it could have formed when Earth was young.

eventually the original chemical is recreated. Diagrams of these cycles look a bit like the map of the London Underground, albeit as drawn by Jackson Pollock during an amphetamine high. The cycling of chemicals may seem pointless, but the intermediate steps produce useful things.

One example is a process used by some bacteria called the reverse citric acid cycle.[8] It uses a molecule of citric acid: the stuff that gives lemons their flavour. The citric acid goes through eight transformations, at the end of which it is recreated. During the cycle, carbon dioxide and water are combined to form simple organic compounds like acetate and oxaloacetate – which can be used to make things like amino acids.

So perhaps what the first life needed was a simple metabolic cycle. Harold Morowitz* suggested this in his 1966 book *Energy Flow in Biology*.[9] He calculated that sending energy flowing through a mix of chemicals would inevitably start some kind of cycle of reactions.[10] In other words, metabolic cycles might get started even in a lifeless world, given the right energy source. 'Molecular organization and material cycles need not be viewed as uniquely biological characteristics; they are general features of all energy flow systems,' he wrote.

This may seem like a rather large claim. Why should pumping energy into a mix of chemicals set up a cycle of chemical reactions? However, analogous processes happen all the time. One example is the water cycle on Earth, which is entirely driven by energy from sunlight. Water from the sea and other water bodies

* We saw in chapter 9 that, in later years, Morowitz became a proponent of the vesicle-first or compartmentalisation-first hypothesis.

evaporates, becoming water vapour that rises into the air. There it cools and condenses back into liquid, forming clouds. The water eventually rains back down, enters rivers and flows back to the sea where it started. The whole system is simultaneously dynamic and static. Trillions of water molecules are always on the move, yet the sea stays roughly the same size, rain is always falling somewhere, and rivers are always flowing. The cycle would stop without the energy from the Sun, but as long as the energy keeps flowing in, the cycle keeps turning.

Still, this was all rather abstract. Morowitz offered no detail on what this simple metabolic cycle might have looked like. It is no wonder that first the primordial soup hypothesis, buoyed by Miller's experiment, and later the RNA World, held sway until the late 1980s. Meanwhile, the idea that life began with a metabolic cycle, a system for absorbing energy and putting it to use, lay moribund. It was finally shocked into life, not by an established researcher in biochemistry, but by a patent lawyer.

Günter Wächtershäuser (pronounced 'Vesh-ters-hoy-zer') was born on the eve of war in the small town of Giessen, Germany, in 1938. Six years later, most of the town was destroyed by Allied bombs. The young Wächtershäuser survived and studied chemistry at the University of Marburg. After obtaining his PhD, he became a patent lawyer specialising in chemical inventions. He is politely irascible, despising sloppy thinking.

One of his tutors was physical chemist Hans Kuhn, who published ideas about the origin of life in the 1970s.[11] Reading Kuhn's work was Wächtershäuser's first exposure to the question and he was fascinated.[12] Kuhn, like many in the 1970s, thought life began with RNA. His twist was that he thought nucleotides would diffuse into porous rock, where they would become trapped and link up to form RNA. The pores inside the rock would act almost like cells, providing a safe home for the RNA.* Kuhn evidently thought he had the problem cracked, and Wächtershäuser saw no particular issue with primordial soup, so he did nothing further. But a seed had been planted.

A decade later, Wächtershäuser met two people who would change the course of his life. One was Carl Woese, who as we saw in chapter 6 figured out how different microbes are related. The pair were introduced in 1982 by George Fox, Woese's collaborator, who had become related to Wächtershäuser after his widowed father married Wächtershäuser's wife's widowed mother. Woese told Wächtershäuser about the problems with the primordial soup hypothesis. 'That became a big question in my mind,' says Wächtershäuser.

That same year, Wächtershäuser also met and befriended Karl Popper, a philosopher of science whose work had fascinated him for years. Popper is many scientists' favourite philosopher, because he tackled the question of why scientific claims should be trusted. He argued that an idea is only scientific if it can be disproved. For instance, if somebody says all sofas are purple, that is a scientific claim: it can be disproved by finding a pink sofa.† Popper saw the history of knowledge as a sequence of ideas that all turned out to be not quite right, forcing people to come up

* The possible importance of rocks with holes in them will recur in a big way in chapter 11. Consider this a tiny little tease of what lies ahead.

† A good example of an un-falsifiable claim would be: 'God exists and loves us, which is why nice things happen – but anything bad that happens is part of his divine plan, which is beyond our comprehension.'

with better ideas. Similarly, the story of evolution is one of life finding solutions to problems, and then improving them when new problems arose. 'If you go back, you arrive at the origin of life, the origin of all problems,' says Wächtershäuser.

After meeting Woese, Wächtershäuser began thinking in earnest about the origin of life, with Popper as a mentor. He began by throwing out everything that had been done so far: the primordial soup, the RNA World and even familiar molecules like amino acids. What was needed was something 'drastically different from anything we know'.

The result was the Iron–Sulphur World hypothesis,[13] which Wächtershäuser first outlined in 1988.[14] The core idea was that the first life needed a source of energy, and Wächtershäuser identified a chemical reaction that he thought might serve: the formation of pyrite. This mineral is famously called 'fool's gold' because its colour resembles gold to the untrained eye, but it does not contain any gold. Instead, it is a crystal of iron disulphide, containing two sulphurs for every iron. Pyrite can be formed by reacting iron with hydrogen sulphide, emitting hydrogen as a by-product. Crucially, this reaction releases electrons, which can be used to react carbon dioxide with hydrogen to make water and formaldehyde: a simple organic chemical from which many others, including amino acids, can be made.

In Wächtershäuser's scenario, as soon as liquid water formed on Earth, some of it would have become laced with iron and hydrogen sulphide seeping out of the hot rocks beneath the surface. 'The earliest organisms are seen as arising in this environment and acquiring their energy by continuously forming and dumping pyrite,' Wächtershäuser wrote.

What would these primordial organisms have looked like?

Wächtershäuser described them in riddling terms. 'These organisms are acellular and lack a mechanism for division, yet they can grow. They possess neither enzymes nor a mechanism for translation [making proteins using instructions from genes], but they do have an autocatalytic metabolism. They do not have nucleic acids or any other template, yet they have inheritance and selection. Although they can barely be called living, they have a capacity for evolution.'

These first organisms were clusters of chemicals attached to the surface of an underwater mineral – possibly pyrite crystals again. Any chemicals that became detached would be swept away into the water and lost for ever, so evolution (for want of a better word) would have favoured those that bonded strongly to the pyrite. Wächtershäuser argued that this favoured larger molecules, and suggested that sugars bonded to phosphates (much like those in nucleic acids) might have formed, along with short chains of amino acids. This layer of interacting molecules was only ever one molecule thick, so it was essentially a chemistry-set version of *Flatland*.*

These molecules could then react with one another, establishing a simple metabolic cycle, powered by the formation of pyrite and living on the surface of yet more pyrite.

It may seem odd to suppose that such a system would form on

* Sadly, not everyone has read Edwin Abbott Abbott's delightful little novel, which conceals a satire of the British social hierarchy in its fascinating speculations about life in a world with only two spatial dimensions instead of the familiar three. Readers who are unfamiliar might instead recall the 2014 *Doctor Who* episode 'Flatline', which features two-dimensional beings called the Boneless that kill people by flattening them, turning them into what seems to be graffiti. It's the same idea, but played for chills instead of dry laughs.

the surface of pyrite. Wächtershäuser was not actually wedded to this specific mineral, but the crucial point is that metals like iron are excellent catalysts. Enzymes often rely on them: at the cores of many protein enzymes are single atoms of iron and similar metals, without which the enzymes don't work. So a good-sized chunk of iron pyrite, or something similar, offered a carpet of catalysts to help the molecules react.

By taking in ever more carbon dioxide and transforming it into more organic molecules, the molecules could spread over more of the mineral surface. As ever more kinds of molecules were added, new reaction cycles would arise, and if the new one worked better than an existing one, it would replace it. Ultimately, this chemical evolution would throw up the molecules needed to make cell walls, DNA and so forth. The first simple cells would form and leave the mineral behind.

To anyone whose knowledge of life is limited to familiar organisms like pigeons and trees, Wächtershäuser's 'precursor organisms' seem bizarre. A cluster of chemicals stuck to the surface of a crystal, living off the chemical reaction between iron and hydrogen sulphide? Such things don't seem lifelike at all. But dig deeper into the world of single-celled organisms, and Wächtershäuser's creations do not seem quite so weird after all.

Five years earlier a microorganism had been discovered that got its energy from the reaction between sulphur and hydrogen.[15] This was *Pyrodictium*, an archaean found living in scalding-hot waters pumping from the seabed in shallow waters off Sicily – aptly, near the island of Vulcano. When Wächtershäuser's friend Karl Stetter and his colleagues grew *Pyrodictium* in their lab, pyrite was formed. Based on this, Wächtershäuser suggested that some single-celled organisms, including *Pyrodictium*, had retained 'the ability to grow by acquiring energy from pyrite formation'. In other words, his idea fitted with how some modern organisms survive.

Indeed, what really fascinates Wächtershäuser is the question of why life uses the chemical reactions it does. 'I am not interested in the origin of life,' he says, unexpectedly. 'The origin of life is something of a mystical distraction.'

There is a second sense in which Wächtershäuser's ideas were radical. Not only had he broken away from primordial soup – disputing everyone from Oparin to Orgel – he was also proposing something controversial about the first metabolism.

While organisms have a bewildering number of ways of getting energy, they all boil down to two basic options. One is to take in energy-rich molecules and break them down to release their energy. That's what we do when we eat food, and it's called heterotrophy. The alternative is to first make the energy-rich molecules yourself, as green plants do when they make sugar by harnessing sunlight. This is autotrophy.

Almost everyone who had tried to imagine the first organism assumed it was a heterotroph, relying on ready-made food from the primordial soup. It just seemed easier: autotrophs require elaborate equipment that heterotrophs do not. For instance, green plants have things called chloroplasts to harness sunlight, which animal cells don't need.

Hardly anyone had argued for an autotrophic origin. One of the few who did was Henry Fairfield Osborn, a palaeontologist famed both for giving *Tyrannosaurus rex* its name and for being publicly racist even by the standards of early twentieth-century America.[16] In his 1916 book *The Origin and Evolution of Life*, Osborn suggested that the first life relied on 'heat energy . . .

derived either from the earth or from the sun or both'.[17] This was an autotrophic origin, but it was muddled: many things can be heated up without coming to life.

In contrast, Wächtershäuser offered cogent arguments. He suspected the primordial soup didn't exist, or at least that there weren't enough nutrients floating in the sea to keep a hungry organism alive. Besides, heterotrophs have to break down their food to extract its energy (and perhaps store it in something like ATP), which also takes plenty of enzymes. Autotrophy, he reasoned, was the simpler option. Energy-releasing chemical reactions were going on all the time, so the key was to find one that could be harnessed to make the molecules of life.

All in all, Wächtershäuser had poked an awful lot of sleeping bears. But he does not remember undue vitriol. 'There were negative reactions, of course,' he says. 'An outsider enters science, and not only enters science, but claims to have an idea that revolutionises this field. You can't really blame them.'

It helped that Wächtershäuser linked his ideas to facts about living microorganisms. In 1990, he went further and drew links between his proposed metabolic cycles and the reverse citric acid cycle – the metabolic process used by some bacteria that was described earlier.[18] This bolstered his case, but many remained deeply sceptical.

Miller, who by this point was very much in the 'criticise anyone who challenges me' phase of his career, was unimpressed. 'Wächtershäuser,' he complained, 'offers no evidence that his schemes are mechanistically plausible, or even possible.'[19] His ideas were 'imaginative and original', but they were also 'ambiguous', and 'none of it is plausible'. 'The time is now ripe,' he concluded, 'for the demonstration of one or two of the many provocative surface-catalyzed chemical syntheses proposed by Wächtershäuser.' In other words, put up or shut up. Wächtershäuser needed experimental evidence.

Wächtershäuser knew full well that his ideas threw up a lot of questions.[20] But he had not done practical experiments for decades, so he needed a partner.

In the early 1990s, he teamed up with Stetter, who showed that pyrite really can be made from hydrogen sulphide – largely as Wächtershäuser had proposed, except that the other reactant was iron sulphide rather than simply iron.[21] Stetter also demonstrated that the formation of pyrite could drive the creation of organic molecules, including amino acids.[22]

Since the mid-1990s, Wächtershäuser has mostly worked with Claudia Huber at the Technical University of Munich. They have produced a series of impressive experimental findings, although they have not constructed anything terribly lifelike.

For example, in their first paper Huber and Wächtershäuser converted carbon monoxide into acetic acid – the chemical found in vinegar – by first making a thioester.[23] In living cells today, thioesters are intermediate steps in the formation of many essential chemicals. 'Thioesters were the holy grail of biochemistry,' says Wächtershäuser. Indeed, cell biologist Christian de Duve devised an entire 'Thioester World' hypothesis for the origin of life.[24] Acetic acid is similarly essential.

Making thioesters and acetic acid was unquestionably significant. But Huber and Wächtershäuser had to tweak things to make it work. Instead of the pyrite-forming reaction, they used a slurry of nickel sulphide and iron sulphide.

The following year they linked amino acids up to make small

proteins, again using a mix of nickel sulphide, iron sulphide and carbon monoxide.[25] Later, they found that this protein-manufacturing reaction could run in parallel with a second set of reactions that broke the proteins back down into amino acids. This was a simple metabolic cycle, constantly generating new mini-proteins and breaking them back down again, powered by a flow of carbon monoxide.[26]

These are impressive demonstrations of what iron and sulphur chemistry can contribute to life's origin. However, many unanswered questions remain. In particular, it has not yet been possible to use iron–sulphur chemistry to power a self-sustaining autocatalytic set. Hence, the reaction cycles created by Huber and Wächtershäuser are arguably not quite as biologically relevant as those demonstrated by nucleic acid and protein devotees – both of whom have created autocatalytic sets. There is also the matter of exactly how something resembling a cell could have been formed.

Still, there is no doubt Wächtershäuser's ideas were brilliant, and they will remain brilliant if they turn out to be wrong. Like Cairns-Smith, he forced everyone to look at the problem of the origin of life from a new perspective. His root idea that life needed an energy source from the get-go is surely correct. Furthermore, while Wächtershäuser may be wrong about iron–sulphur reactions being the energy source, he could well be right that the first energy source was chemical. The only obvious alternative is light energy, and in the history of evolution the ability to harness light emerged after the ability to feed off chemicals.

Perhaps the biggest compliment that can be paid to Wächtershäuser is that a competing theory emerged within years of his. This new idea was also metabolism-first, emphasising the need for chemical energy to get life started, and at first glance it can be easy to get the two muddled up. But in almost all its details the new hypothesis was radically different. Its story begins in 1977, in a submersible hundreds of metres below the surface of the Pacific Ocean.

Chapter 11

Born in the Depths

On 8 February 1977, a research vessel called the R/V *Knorr* cruised out of the western end of the Panama Canal and set out into the Pacific Ocean. Its destination was the East Pacific Rise, a ridge on the seabed along the boundary between two tectonic plates. Previous expeditions had found plumes of warm water above the Rise, and bottom-dwelling fish had been found floating dead near the surface.[1] This led oceanographers to suspect that the Rise was the site of 'hydrothermal vents': a phenomenon never before seen, where hot water, heated by volcanic rocks and magma beneath the sea floor, surged up into the cold ocean.

The *Knorr* was carrying a team of oceanographers that included John 'Jack' Corliss, Tjeerd van Andel and Robert Ballard.[2] Their task was to find the hydrothermal vents, if any truly existed. None of them knew it, but what they found would revolutionise biology – and ultimately lead to a new, metabolism-first hypothesis for how life began.

From above the Rise, the team lowered a cage full of cameras, strobe lights and sensors, known as ANGUS,* into the water. Attached to the ship by a cable, ANGUS was towed along 4.5m above the sea floor, taking photos every ten seconds. Around midnight, it registered a brief spike in temperature: it had passed through a plume of warm water. After ANGUS ran out of film, it was hauled up to the surface and all 70,000 photos were developed the following day.

Most of the photos showed barren rock: a seemingly endless expanse of cooled lava vomited out of the seabed. But the thirteen photos taken during the temperature anomaly were unrecognisably different. They showed thriving life. 'The lava flow was covered with hundreds of white clams and brown mussel shells,' Ballard later wrote.[3] 'This dense accumulation, never seen before in the deep sea, quickly appeared through a cloud of misty blue water and then disappeared from view.' This was a complete surprise. The team did not include a single biologist: they had not expected to need one.

The next day, a second ship arrived. It was carrying a submersible called DSV *Alvin*, which had once been used to locate a hydrogen bomb that the US Air Force had lost at sea. In 1986, Ballard would take *Alvin* deep into the North Atlantic to explore the wreck of the *Titanic*. However, this time Corliss and van Andel rode in *Alvin*, with pilot Jack Donnelly. Soon after sunrise, *Alvin* gradually sank to a depth of 2700m. The water temperature was just 2°C, the pressure was crushing, and it was pitch-dark. *Alvin*'s lights illuminated a tiny patch of sea floor,

* Short for 'Acoustically Navigated Geophysical Underwater System'. It must take a lot of effort to make these acronyms work.

surrounded by Stygian blackness. Everywhere was barren, cold rock, until the submersible reached the co-ordinates. The temperature suddenly rose to 8°C, and Corliss and van Andel found themselves staring at hydrothermal vents.[4]

Streamers of shimmering warm water were flowing upwards from cracks in the lava. As they mingled with the cold Pacific water they turned cloudy blue: the chemicals dissolved in the vent water were coming out of solution and forming clouds of metallic dust. This was the deep-sea equivalent of the hot springs found on land, where water underground is heated by the surrounding rocks then surges up to the surface. Think of the simmering ponds where Japanese macaques bathe to keep warm, surrounded by snow, and now imagine the same hot, chemical-rich water instead welling up from the sea floor into dark, frigid salt water.

The vents were crawling with life: not just the clams and mussels from the photos, but scuttling crabs, a purple octopus and fields of tubeworms half a metre long. When the team brought samples to the surface, they got another shock: the animals stank of rotten eggs. The whole area was rife with hydrogen sulphide. They did not have facilities to preserve the specimens properly, but they managed to keep some by appropriating a bottle of vodka intended for other purposes.

Later expeditions discovered that the vents come in several forms. Those that are directly above the hottest rocks can blast out water at temperatures of 380°C. The minerals from the vent water instantly crystallise, forming tall chimneys blackened by sulphides. These super-hot vents are called 'black smokers'.[5]

The immediate question was how the animals around the vents could survive. Before the 1977 expedition, biologists had assumed the ocean floor was a kind of desert where nothing could live: not because of the cold and pressure, but because of the darkness. On land, all life depends on sunlight. Plants and various bacteria use the Sun's energy to convert carbon dioxide and water into sugars, which they use as food. Animals then eat the plants, and are eaten by other animals. When those animals die, they are eaten by decay organisms like fungi, enriching the soil and allowing plants to grow. Without the plants to start the cycle by harnessing sunlight, the ecosystem would collapse. The vent organisms had no sunlight, so should not have existed.

The crucial clue was the stink of hydrogen sulphide: the chemical Wächtershäuser would later propose as the energy source for the first life. It transpired that the tubeworms on the vents had crystals of sulphur in their guts. A young Harvard graduate student named Colleen Cavanaugh heard about this and realised there must be bacteria living in the worms' guts, which obtained energy by converting hydrogen sulphide into sulphur. Cavanaugh guessed that this was the basis of the entire ecosystem.[6]

Within a few years, the world learned that hydrothermal vents were real and supported ecosystems quite alien to us. The vent organisms were not powered by sunlight, but by chemicals like hydrogen sulphide pulsing from the sea floor. They thrived in darkness, in an environment where the water temperature varied from near-freezing to scorching hot within metres. By any measure, the vents were an astounding discovery.

But for Jack Corliss they represented something more. He soon came to suspect that they were the site of life's origin. With two colleagues, he argued that the underground chambers below the vents were 'ideal reactors for abiotic synthesis'.[7] Simple chemicals like carbon dioxide, ammonia and hydrogen would emerge from the hot rocks and react in the ways people like Miller had described, producing amino acids and later larger molecules like proteins. Corliss noted that montmorillonite clays* are common in hydrothermal systems, and would have speeded up many of these reactions. As the biological chemicals moved up towards the sea, the water would cool and more intricate structures would form, until simple cells emerged.

At this point, it should be no surprise that Stanley Miller was having none of it. In 1988 he and Jeffrey Bada published a strongly worded rebuttal.[8] Their argument was simple: the scorching temperatures in the vents would destroy any biological molecules that formed. This was no mere assertion: they had performed experiments showing that amino acids broke down within twenty minutes at temperatures of 250°C. It followed that any amino acids that formed below vents would have to hurtle upwards to escape the hot zone before they were destroyed. Similarly, sugars would survive 'seconds at most' and proteins would shatter into amino acids. The whole idea, they said, 'can be dismissed'.

The attack worked: Corliss's idea fell by the wayside. But a British geologist named Mike Russell thought there was something there worth salvaging. Scorching-hot vents like those on the East Pacific Rise would never do: Miller and Bada had shown that. But there might be less extreme vents that were more hospitable.

Michael John Russell occupies a curious position. Among scientists who study the origin of life, he is famous as the person

* Yes, that one again.

behind one of the most prominent hypotheses. Unlike the RNA World, which was shaped by dozens of researchers, the alkaline vent hypothesis was largely Russell's baby. Yet he has no Wikipedia page ('too fucking busy', he says) and rarely gives interviews. He turned eighty in 2019 but remains sharp and fiery, quoting freely from Rainer Maria Rilke and Bob Dylan. He sees himself as a modern Galileo, fighting for truth against a hidebound scientific establishment.

His path to his chosen field was circuitous.[9] After leaving school in 1958, he got a job making aspirin in a factory just outside London. By studying in the evenings he got a degree in geology and chemistry, and five years later arrived, as a volunteer geologist, in the Solomon Islands in the Pacific Ocean. 'It was the last bastion of the British Empire,' he recalls. 'All the imperialists that couldn't get jobs in Canada or Australia or the US went to the Solomon Islands, so you can imagine they were the bottom of the pit. Horrible people.'

Russell eventually entered academic research. By the early 1980s he was working at the University of Strathclyde and regularly exploring mineral deposits near Silvermines in Ireland. The village took its name from rich metal deposits nearby, including a whitish mineral called barite. There he found the first clue that would lead him towards his big idea.

Russell had long suspected, encouraged by a former mentor, that such mineral deposits were traces of ancient hydrothermal vents. When he saw the photos of black chimneys on hydrothermal vents, he wondered if such structures might be preserved in the barite. He looked, and found tubes of pyrite around one centimetre in diameter.[10] This, he argued, was evidence that the barite and other metallic minerals near

Silvermines had once been spewed out of a hydrothermal vent.

But these ancient vents were not like the screamingly hot black smokers, where temperatures could reach 400°C. Russell was convinced that they did not even reach 150°C, so while hardly cold they were less hostile to biological molecules. There would have been plenty of reactive chemicals, but the environment was not so extreme that products like amino acids would be obliterated. Furthermore, the little tubes of pyrite might have acted as 'culture chambers', safe havens where the molecules of life could build up and interact.

From 1983 onwards Russell and his colleague Allan Hall mused on these ideas, but published nothing. Instead, much of Russell's time was spent fighting unsuccessfully to save his university's applied geology department from closure. Margaret Thatcher was in power in the UK, spending cuts were in fashion, and university departments were being scythed.

He finally spoke up when Miller and Bada attacked Corliss's vent hypothesis. 'It didn't occur to me it was the same Miller,' Russell says. He and Hall fired off a reply, aided by two colleagues including Graham Cairns-Smith, in which they argued that their more equable vents were worth a look.[11]

It was 1988, the same year Wächtershäuser proposed his Iron–Sulphur World based on pyrite chemistry. Russell's ideas at this time also revolved around pyrite, and at first he drew strong connections between Wächtershäuser's ideas and his.[12] He also offered to collaborate, but this never panned out and the two have never got on. Instead, Russell soon broke out into new ground.

He was inspired, in part, by a French medical doctor named Stéphane Leduc, who was active in the late 1800s and early 1900s. Leduc tried to show that simple living things could form due to purely physical processes. By mixing certain chemicals in glass jars, he created beautiful and lifelike patterns, including some that resembled clusters of cells. Importantly for Russell, he was a pioneer of metabolism-first, arguing that 'the essential phenomenon of life is nutrition'.[13] However, Leduc struggled to publish in France, because the Academy of Sciences thought his ideas sounded like the discredited notion of spontaneous generation (see chapter 1). Nowadays his experiments are little more than a curiosity, because their resemblance to living things is superficial. Nevertheless, he was an inspiration.

Russell's first step was to recreate the fossil pyrite tubes in the lab.[14] He dissolved sodium disulphide in water and passed it through a narrow hole, on the other side of which was salt water laced with iron dichloride. Where the two solutions met, a layer of jelly formed. Bubbles and blobs emerged, broke off and became thin vertical tubes a few millimetres across and several centimetres high. These tubes, Russell concluded, were what emerged from his proposed alternative vent. Not a hard rock, but a wobbly jelly rich in sulphides and metals; a jelly inside which other chemicals could gather.

In 1989, Russell took a crucial step. He decided that his vents were not just cooler than black smokers, they were chemically different. Their water was alkaline, whereas the black smokers' was acidic. Most of us think of acids as corrosive liquids that dissolve metal, like the acid blood of the creature in *Alien*,* but

* The creature is not called a 'xenomorph', despite what the internet would have you believe. Xenomorph is a generic word used *in the sequel* to mean 'unidentified alien'. This pedantic correction has no relevance whatsoever, but that's what footnotes are for.

actually an alkali can be just as corrosive. At the chemical level, acids are molecules that shed protons – the nuclei of hydrogen atoms – when they are dissolved in water. Alkaline chemicals do the opposite and absorb protons. Chemicals with no tendency either way are 'neutral'.

The difference in the vents' water chemistry, Russell believed, was due to their location. While the black smokers are powered by molten magma heating water, his vents were driven by a chemical reaction between solid rock and water. This reaction is called serpentinisation, because it forms a beautiful green mineral called serpentinite which looks like snake scales. Serpentinisation releases heat, so Russell argued that serpentinisation below the seabed would result in warm and alkaline waters rising through the layers of rock and out into the open water.

It may not seem obvious why the vent water being alkaline instead of acid would be good, if both are corrosive. However, when Earth was young the ocean may have been acidic, because some of the carbon dioxide in the air would have dissolved in it, forming carbonic acid.* If sulphide blobs like the ones Russell made had indeed formed on the alkaline vents, the water in them would be alkaline while that outside was acid. This meant the water outside had more protons in it than the water inside: there was a proton 'gradient'. The protons outside the blobs would 'want' to flow into them but would struggle to do so.

This may still seem arcane, but proton gradients are central to metabolism and arguably to life itself. To understand why,

* Today the seas are slightly alkaline, but they are rapidly acidifying because of our emissions of carbon dioxide.

we have to pop back in time several decades and meet one of twentieth-century science's most intriguing figures.

Peter Mitchell was born into privilege in south-east England.[15] His father was a civil servant who received an OBE and his uncle was president of construction firm George Wimpey (now Taylor Wimpey). In 1939, Mitchell went to the University of Cambridge, where despite being considered 'the bright one' he achieved indifferent grades. He eventually scraped a PhD in biochemistry in 1951. By one account, he was 'especially gifted and imaginative' but had an 'inability to get his thinking across to others'.[16]

While working on his PhD, Mitchell began collaborating with biochemist Jennifer Moyle, who became his closest colleague for three decades. Moyle had also arrived at Cambridge in 1939: a time when the university did not award degrees to women but merely gave them the 'title of a degree'. She spent most of the war in military intelligence before returning to biochemistry. The pair were a perfect match: Mitchell was an ideas man, Moyle a rigorous experimenter and his confidante.

However, in 1955 Mitchell left Cambridge for Edinburgh. This was partly to advance his career, but his personal life had also become tangled. His first marriage to Eileen Rollo in 1944 had produced two children, but the couple divorced in 1954. Mitchell had embarked on an affair with a woman named Helen Robertson, who was married with two small children. Forced by her husband Pat to choose between Mitchell, whom she adored, and her children, Helen moved with her husband and children to Bristol. Undeterred, Mitchell visited her in disguise after sending her love letters hidden inside egg cartons. But Helen could not handle being pulled two ways and ended the affair by letter, writing: 'Dearest Pete, I can't do two things at once and therefore mustn't ever see you again. Good-bye for ever. Love, Helen.'

So it was that Mitchell set himself up in a basement room at the University of Edinburgh. He invited Moyle to join him and they began establishing themselves. Soon after, Mitchell was invited to lecture in Bristol and managed to visit Helen. Their love had not dwindled, and she ultimately left her husband, taking the children with her. Peter and Helen finally married in 1958 and remained together until his death in 1992.

It should be clear from the saga of Mitchell's love for his second wife that he was a deeply passionate, confident and flamboyant person. He would need all those faculties now. In his eight years at Edinburgh, he and Moyle would devise their most important idea – an idea they would have to defend for years to come.

Mitchell's obsession was adenosine triphosphate (ATP), the molecule all cells use to store energy. Metabolic cycles (like the ones Wächtershäuser explored) obtained energy for the cell, and stored this energy in ATP. The cells then broke down the ATP when they needed the energy. The question was, how did the cells make the ATP? How was that energy harnessed? In essence, life is battery-powered, and Mitchell was asking how the batteries were charged.

Part of the answer was clear. The metabolic cycles released electrons, and these travelled down a series of proteins embedded in a membrane in the cell. As each electron moved from one protein to the next, it released some energy and this was used to make ATP. It's a bit like a football bouncing down a staircase, with each impact releasing energy. Astrobiologist Charles

Cockell has argued that this process of pass-the-electron is fundamental to life, because electrons are the most accessible bits of atoms, as they're on the outside.[17]

The problem was, nobody knew how the energy from the electrons was physically used to make ATP. Mitchell's solution, developed over several years and finally stated in print in 1961, struck most of his colleagues as bizarre.[18] He supposed that there was an enzyme for making ATP, which is simple enough, and that it was also embedded in the same membrane within the cell. However, Mitchell also imagined that there was a proton gradient in place, with lots of protons on one side of the membrane and few on the other. This was what the movement of the electrons achieved: every time a protein in the membrane received an electron, it would pump a proton across the membrane so that they accumulated. The protons would not be able to pass back through the membrane itself, but they could travel through the ATP-making enzyme – giving it the necessary energy to make ATP.

This 'chemiosmotic hypothesis', as Mitchell called it, was a brilliant stroke of inspiration. It is also weirdly overcomplicated, as if someone had built a power station (the moving electrons) but then used the resulting electricity to pump water uphill (the proton gradient), then let it flow back downhill to drive a turbine (the ATP-making enzyme). Why not just use the energy from the power station? Mitchell and Moyle duly spent the next few years performing supportive experiments, while fending off criticism from colleagues convinced that the process could not be so unwieldy.[19] Their precarious position was made worse in 1962 when Mitchell became seriously ill with gastric ulcers. After he refused his surgeons' suggestion that they remove 80 per cent of his stomach, he was told that he must resign his position at the university and convalesce.

This would have finished most people off, but Mitchell was resourceful and had serious money. The previous year, he had bought a holiday home: a small cottage that stood at the entrance to the grounds of a manor building called Glynn House in Cornwall, south-west England. He chose it mostly because it was as far from Edinburgh, the climate of which he had come to detest, as it is possible to get without leaving Britain. Then in 1962 he bought Glynn House itself for just over £2800 – about £60,000 in today's money.

Today Glynn House is a listed building, but at the time it was riddled with dry rot. Mitchell had noted that 'only a lunatic would want to become involved with it', but he did so anyway. He spent the summer of that year working on it, putting up sheets of polythene to keep the rain out. At the time he was simply planning 'to save a fine old building from destruction'.

Then the ulcers struck, and Mitchell and Helen retreated to the cottage. Without an academic position, Mitchell had the idea of restoring Glynn House and converting it into an independent research institution. He hadn't much liked traditional academia anyway. Moyle was willing, so they spent the next few years fixing up the building, using £70,000 of Mitchell's money (about £1.5 million today). Mitchell had also taken on eight Jersey cows that lived on the estate and needed milking by hand twice a day. Once the building was shipshape, Mitchell set up a small research establishment called Glynn Research Ltd using yet more money. There, he, Moyle and a few colleagues continued working on proton gradients.

By 1978 the debate had been settled: Mitchell was right, and he duly won the Nobel Prize in Chemistry.[20] Moyle was not awarded a share, and retired a few years later. Mitchell evidently felt guilty about this, because he tried to secure a public honour for her, arguing that 'I know no other biochemist whose experimental skill and judgement is superior'. He also tried to secure her an honorary degree, to make up for Cambridge's reluctance to award a proper one. But he failed in both efforts.

Nevertheless, Mitchell and Moyle had established that proton gradients play a central role in life, providing the energy that drives the manufacture of ATP. This explains why Mike Russell was so enthusiastic about the idea of alkaline vents, where a proton gradient would arise naturally. It offered a source of energy for the first life, and for chemical cycles along the lines of the ones Wächtershäuser had proposed. The machinery for pumping protons over a membrane is complicated and presumably only arose later.[21] But in the vents the protons were in a sense pre-pumped, so life simply needed to harness the energy of the proton gradient. This would also explain why life uses such a convoluted mechanism: it started with the proton gradient, and built the rest around it.

Russell and his colleagues set this out between 1989 and 1993.[22] Unlike Miller's epochal study forty years before, Russell's appeared in small journals and the press ignored them. But they were arguably just as important. Russell had taken several seemingly unrelated ideas – Wächtershäuser's metabolic cycles and iron sulphides, hydrothermal vents, Mitchell's proton gradients – and knitted them together into a compellingly intricate pattern.

Was it true? Russell promoted his ideas incessantly throughout the 1990s, but there were several unanswered questions.[23] The first was simple: do alkaline vents really exist? If they did, could their natural proton gradient be harnessed to make organic molecules like amino acids, without enzymes? And could this lead to the formation of simple cells?

In December 2000, the first question was answered. Scientists aboard a research vessel named *Atlantis* were exploring the Atlantis Massif: a dome on the bottom of the Atlantic Ocean about fifteen kilometres across and almost four kilometres tall. The massif is near the Mid-Atlantic Ridge, the line running north–south through the Atlantic where two tectonic plates are moving slowly apart. The scientists lowered cameras into the cold water and watched a live feed. To their surprise, 'strange-looking snow-white deposits and pinnacles came in and out of view'.[24]

A team led by oceanographer Deborah Kelley went down in *Alvin* (making its second appearance this chapter) and discovered a 'forest of stunning, tall white chimneys' up to sixty metres tall.[25] The researchers onboard were reminded of Greek and Roman columns, and both the massif and ship were called Atlantis, so they dubbed the site 'Lost City'. The name is apt. The rock spires are eerily still and look bleached. There is something deathly and haunting about them. Some look rather like termite mounds, but there is little visible activity. Although there are many microorganisms in Lost City, Kelley's team found few larger organisms, noting only 'a few crabs, sea urchins, and abundant sponges and corals'.

Lost City is a field of hydrothermal vents. The water flowing from the white chimneys is mild compared to that of the Pacific vents, with a temperature between 40 and 90°C. It is also gently alkaline, its pH around 10. The whole thing is powered by serpentinisation, the chemical reaction between rock and water, beneath the seabed.[26] Russell had predicted the existence of such vents over a decade before, and he had been right. A key plank of his theory was correct.

Not that Russell got it all right. He had envisioned bubbles and spires of iron sulphide, but the towers of Lost City are not iron sulphide. They are mostly carbonate minerals, similar to the limestone of the White Cliffs of Dover – explaining their pale appearance. However, this required only a small shift in thinking. These minerals are porous: they are filled with holes, like a sponge. It was easy to imagine these tiny water-filled bubbles in the rock hosting the first life: the first, non-biological cells. Experiments suggest any biological chemicals that formed would gather into concentrated clumps due to the constant flow of liquid.[27]

Around the time of the discovery of Lost City, Russell had also found a new collaborator: the pugnacious and ingenious microbiologist Bill Martin. From Maryland, he spent much of his twenties working as a carpenter, before moving to Germany and obtaining a degree at the age of twenty-eight. During his career Martin has devised a cavalcade of counterintuitive ideas about the early evolution of life, many of which have held up remarkably well. However, his collaboration with Russell only lasted a decade. They disagreed about the chemistry and both were large personalities: Russell compares them to John Lennon and Paul McCartney, who also became unable to work together. Still, Russell is clear that Martin's role was crucial. 'I could not have done without him,' he says. Martin was the perfect partner because he had expertise in microbiology, in which the geologist Russell was an amateur.

Martin's biggest contribution is arguably the suggestion

that the first life used a metabolic system known as the Wood–Ljungdahl pathway.[28] This pathway is used by many bacteria and archaea. It combines carbon dioxide and hydrogen with a larger molecule called coenzyme A, producing a substance called acetyl coenzyme A and some water. Acetyl coenzyme A crops up everywhere in biochemistry, so making it opens up a host of possibilities. For Martin, this was a more plausible starter metabolism than the reverse citric acid cycle proposed by Wächtershäuser, because it does not involve anything so complex as a cycle, and can work in several different ways.

This version of the alkaline vent theory is now one of the most widely cited and admired suggestions for how life might have started. It has been promoted by biochemist and science writer Nick Lane in *The Vital Question*,[29] and even made it to prime-time television in Brian Cox's *Forces of Nature*. Martin likes to point out that it is also regularly discussed in textbooks of biochemistry and cell biology, because it fits with the metabolisms of bacteria and archaea.

Astute readers may sense a 'but' coming, and it is a big one. Go back over the chapter so far and count the number of times you see the word 'experiment'. There won't be many. After Russell's early experiments with iron sulphide membranes, most of the rest of his work has been theoretical – and latterly, with Martin, drawing on existing biochemistry. The whole elaborate edifice has very little experimental evidence behind it.

The first unanswered question is whether organic molecules can form and survive in alkaline vents. Nick Lane's team tackled this question by building an 'origin-of-life reactor' to simulate the interior of a vent.[30] It is a glass cylinder ten centimetres across and tall, with tubes plugged into it through which liquid can be injected. By filling the reactor with acidic liquid to mimic the primordial ocean, then adding alkaline 'vent' fluid, they created hollow tubes and spires. In this tabletop alkaline vent, they were able to convert carbon dioxide into formaldehyde. This could in turn be converted into sugars, including ribose and deoxyribose from RNA and DNA respectively. However, this latter step only worked when they supplied concentrated formaldehyde, not the minute amounts that actually formed. This is an old prebiotic chemist's trick, and not a plausible one.

It should be said that Lane's team made no big claims, calling the study 'a preliminary proof of concept'. But since they first described the reactor in 2014 they have published nothing further about it. Furthermore, Wächtershäuser has pointed out that the amount of formaldehyde produced was so small that it could have been contamination – a problem that has cropped up before, but one which the team did not properly account for.[31]

Experiments by other groups have yielded similarly unsatisfying results. For example, NASA researcher Laurie Barge made an amino acid from a chemical named pyruvate, using a mixture intended to simulate an alkaline vent.[32] Again, this is neat, but so many decades after Miller's experiment, making a handful of bio-molecules is no longer impressive. It is particularly underwhelming in the knowledge that individual vent chimneys are only active for about 100 years, which is not much time for an entire cell to form.[33]

Chemists who study the origin of life are often openly contemptuous of the alkaline vent hypothesis. John Sutherland, who we'll meet in chapter 14, once wrote that the idea 'should, like the vents themselves, remain "In the deep bosom of the ocean buried"'.[34] A common complaint is that Russell and his cohort are not chemists. Again and again, vent sceptics have pointed out that RNA and other key chemicals are unstable in water, because the water attacks them. Russell's supporters argue that the unusual conditions in the vents, coupled with the steady flow of new chemicals from below the seabed, should counteract this problem. It is only fair to say that this has not been shown experimentally: it is also fair to say that it has also not been disproven, because so few experiments have been done.

However, the biggest problem for the alkaline vent hypothesis is its most unique element, which at first sight seems the most convincing: the idea that a natural proton gradient could supply the energy to kick-start metabolism. This idea is a brilliant intuitive leap, but there is no experimental evidence. All life does use proton gradients, but all life also uses ribosomes and nobody thinks ribosomes were present at the very beginning.

The problem is twofold. First, we do not know that there are sharp proton gradients within alkaline vents like Lost City.[35] Instead, alkali may slowly blend into acid over the length of each chimney, in which case the proton gradient will be too gentle to generate useful power. Second, the enzymes that life uses, including the one that makes ATP, are big and complex. So far, nobody has found a simpler version that works and could plausibly have formed. This absence is glaring, just as the lack of a self-replicating RNA has been a problem for the RNA World.

In the last few years Russell has tried to solve this problem. It seems unlikely that the first life used ATP itself, as the adenosine part of the molecule is elaborate. However, the key part is the chain of phosphates, and these 'polyphosphate' chains may simply have formed on their own. Indeed, Harold Morowitz pointed out in 1992 that many microorganisms make polyphosphates and used them to store chemical energy.[36] Russell now suspects that the first life used the simplest possible polyphosphate: pyrophosphate, which is simply two phosphates strung together.

To integrate pyrophosphate into his scheme, he has abandoned the idea of iron sulphide bubbles. 'There were a lot of people who loved that because it looked like a cell,' he says. However, he now thinks the pores in the rocks of alkaline vents were lined with multiple thin layers of 'green rust'.[37] Most of us have seen green rust, for instance on old iron ships that have long been exposed to seawater. It is a compound of iron, hydrogen, oxygen and other chemicals, and Russell found it often formed when he tried to simulate conditions in the vents. He posits that these layers of green rust within the rock pores were the first cell membranes.

This may seem an odd addition – surely the pores themselves were suitable containers for proto-life? – but Russell thinks the film of green rust could have been how the first life harnessed proton gradients to make pyrophosphate, without an enzyme. His proposal is that the proton gradient over the green rust membrane pulled phosphates and protons into narrow gaps in the green rust crystals, where they fused to form pyrophosphate. This was then released into the gaps between green rust layers, where it drove the synthesis of other biological molecules. It is an ingenious idea, which he is now trying to test. 'If we can't show in three years that that works, then we're in dead trouble,' he says.

However, Russell is facing a new obstacle. In 2019 he lost his long-standing position at NASA's Jet Propulsion Laboratory, and he is now living in Italy. He's trying to get the experiments done at European Universities.

Meanwhile, genetics has yielded a startlingly powerful piece of evidence in favour of the alkaline vent hypothesis. In 2016, Martin's team published a detailed reconstruction of the Last Universal Common Ancestor (LUCA) from which all modern organisms are descended. They did so by examining the genes of 1930 microorganisms, searching for genes that they all shared – which probably existed in LUCA. This was not easy, because microbes sometimes take a gene from an unrelated species; a process called horizontal gene transfer. This can make a gene appear universal and ancestral, when it actually evolved later and then spread. After the team had cleaned up the data as best they could, they were left with 355 gene families that seemingly existed in LUCA.[38] These suggested that LUCA lived somewhere hot – which is compatible with an alkaline vent but doesn't prove it – and that it used the Wood–Ljungdahl pathway to make biological molecules, as Martin predicted. Furthermore, it seems LUCA had the equipment to harness a proton gradient, but not to generate one – which fits with the idea that it relied on a natural proton gradient in a vent. This latter finding is striking, but must be taken with a pinch of salt because of the horizontal gene transfer problem.

The alkaline vent hypothesis is beautiful and detailed, and lines up with microbiology. But that doesn't make it true. Plenty of beautiful, plausible-seeming ideas have turned out to be wrong. It is not at all clear that the hypothesis can surmount its many problems.

However, several of its key elements are so compelling that the true theory must surely incorporate them, or find some other solution to the problems they address. A source of chemical energy to fuel metabolism is obviously crucial, but possibly so is the ability to harness, or even generate, a proton gradient.

Finally, it is striking that the hypothesis attempts to make two of the components of life – a metabolic cycle and a compartment – at once. It is this more holistic approach that is arguably most significant. Rather than trying to do everything with RNA, or with proteins, Russell's idea endeavours to build something that looks more like a complete cell. In this, if nothing else, it hints at a better explanation for life on Earth. In the twenty-first century, many researchers followed Russell's example and stopped trying to do everything with one kind of chemical. Instead, they would find ways to create all of life's components at once. Even if Russell's hypothesis turns out to be wrong, his work clearly foreshadowed this new approach.

PART FOUR

Reunification

'This is a cell. Like all cells it is born from an existing cell. By extension, all cells were ultimately born from one cell: a single organism alone on planet Earth, perhaps alone in the universe, about four billion years ago.'

Annihilation, screenplay by Alex Garland,
based on the novel by Jeff VanderMeer

Chapter 12

Mirrors

We have seen how, in the wake of the discovery of life's inner complexity, researchers devised several new hypotheses for how life began. Each focused on one function, or one key component of life, assuming that it came first and that the rest of the living cell assembled around it. We have also seen that these ideas do not really work. From Fox's proteins-first hypothesis to the RNA World, these simple systems never become terribly lifelike. Instead, a new approach is needed: a way for all the components of life to form together. The alkaline vent hypothesis hints at this new direction, though it has problems of its own.

Some of the first clues to this new approach emerged in the 1990s, when biochemists finally tackled a fiendish chemical puzzle – one that has always been intertwined with the question of life's origin. The problem emerged in the 1800s and was clear to early origin-of-life experimenters like Stanley Miller and Leslie Orgel. But hardly any experimental solutions were attempted. Not until the 1990s was real progress made. The problem lurked in the shadows, even as the RNA World and alkaline vent hypotheses came to prominence and the Oparin–Haldane primordial soup dissolved away.

The problem was this. Every nucleotide, and almost every amino acid, comes in two forms. These variants are mirror images of each other, much like a person's left and right hands. If these chemicals are allowed to form naturally, the result is an exactly equal mixture of both forms. But the processes of life seem to require purity: only one kind of molecule can be used. This seems, at first glance, to be an insurmountable paradox. However, when solutions emerged they pointed the way towards a new account of life's origin.

The first hint of this mirror-image molecule problem was uncovered by French physicist Jean-Baptiste Biot. He came to prominence in 1803, aged twenty-nine, when he reported that stones that had fallen on a small French town were from space, kick-starting the study of meteorites.[1] A decade later, Biot was focused on optics: the study of light.

His great interest was polarised light, which behaves differently to normal light. When light emerges from a source like a lamp and travels towards your eyes, it is moving in a wave – like the wave that can be sent down a taut string if you wiggle the end. Normally, the light waves are wiggling in all directions: up and down, side to side. However, in polarised light the waves all move in the same plane. It is as if the waves have been carefully turned so they line up neatly.

In an 1815 study, Biot shone polarised light through several substances, including sugar dissolved in water.[2] He found the light was affected in a peculiar way. The polarised waves were rotated either clockwise or anticlockwise as they passed through the sugar water and some of the other substances. Somehow, the sugar molecules were twisting the light rays. That was odd, but the inconsistent direction was odder: why would the light rays sometimes rotate clockwise and sometimes anticlockwise? It was as if there were two forms of the sugar, with differing effects on the light, but otherwise identical in their appearance and behaviour.

In the 1840s, Louis Pasteur took the next step – a decade before his debate with Pouchet over spontaneous generation (see chapter 1).[3] He studied a version of a chemical called tartrate, which could also rotate polarised light. Pasteur made crystals of it and carefully examined them. He found there were two kinds of crystal, on which some of the faces were turned either to the left or the right. Although the crystals had the exact same number of faces, there was no way to turn them so that they matched. It was like placing a person's left hand on their right hand: the two never lined up. Pasteur concluded that there were two kinds of tartrate molecule, which were somehow left-handed and right-handed.

However, Pasteur was unable to explain exactly what was going on.[4] At this time, chemists did not understand the shapes of molecules: the structures of DNA and the like would not be known for a century. Even the idea that matter was made of atoms was disputed.

It took another quarter of a century for the solution to be understood. Two young scientists independently alighted on the correct answer and published their ideas in 1874. One was Joseph Achille Le Bel, a twenty-seven-year-old French chemist. The other was the fabulously named Jacobus Henricus van 't Hoff, Jr, a Dutch chemist who was just twenty-two years old and had not even finished his doctorate. In 1901 he would become the first winner of the Nobel Prize in Chemistry.[5]

Both Le Bel and van 't Hoff saw that the behaviour of carbon atoms in molecules was crucial.[6] They realised that each carbon could form bonds with up to four other atoms. In the simplest case, all four would be the same – all hydrogens, say – and there would be no left-handed or right-handed form. Indeed, a single carbon atom surrounded by four hydrogens makes a molecule of methane, which does not rotate polarised light. You can see this by drawing a molecule of methane on thin paper using a strong pen. If you turn the paper over, the image showing through will be identical.

Now let's consider a more complicated molecule: glycine, the simplest amino acid. The central carbon atom carries a carboxyl group, an amino group and, crucially, two hydrogens. Despite glycine's greater complexity, there is still only one version – no right-handed and left-handed forms, and no effect on polarised light – because of the two hydrogens. If you have two molecules of glycine, it is always possible to spin them around so that they match, and if you repeat the trick with the thin paper, you can still get the image to match.

But now consider alanine, a larger amino acid. Here, one of those hydrogens has been replaced with a carbon atom surrounded by hydrogens. This means alanine's central carbon is surrounded by four different things, and the experiment with the thin paper will not work. When you turn the paper over, two of the surrounding groups will always swap places, relative to where they were when the paper was the right way up.

Molecules like alanine that exist in these subtly different forms are said to be 'chiral' (the 'ch' is actually a hard 'ck' sound, and the word rhymes with 'spiral'). Apart from their effect on polarised light, they are almost identical: they look the same, melt at similar temperatures, and undergo the same chemical reactions. These alternate versions are often dubbed left- and right-handed.*

In living organisms, amino acids are always left-handed and nucleotides are always right-handed. It is not clear that this specific arrangement is essential. Maybe life could have instead gone for right-handed amino acids and left-handed nucleotides, or right-handed everything. Rather, it may be like the rules countries have for driving on one side of the road: it doesn't matter which option a country picks, so long as everyone there sticks to it. Certainly, if you give an organism the wrong kind of amino acids or nucleotides, it will hit problems. For example, the beautiful double-helix structure of DNA can only form properly if all the nucleotides are right-handed. A single left-handed one will deform it, making it harder or even impossible for the sequence of bases to be read. Similarly, it is difficult to extend a strand of right-handed RNA if left-handed nucleotides are present, as they gum up the works.[7]

For origin-of-life researchers, this meant yet another problem. It was not enough for molecules like nucleotides to have formed

* Chiral molecules can also be labelled in a different way, which turns on whether they rotate polarised light clockwise or anticlockwise. If it's clockwise, they are 'd' and if anticlockwise 'l', giving d-alanine or l-alanine. This came about because clockwise rotation is also known as 'dextrorotation' while anticlockwise rotation is 'levorotation', the prefixes being derived from Latin. While we're on terminology, the different forms of a chiral substance are called 'enantiomers'. A mixture with equal proportions of all the enantiomers is said to be 'racemic', while a substance solely made up of one enantiomer is either 'enantiopure' or 'homochiral'. I will be avoiding these terms entirely, but they are spelled out here for readers who want to delve deeper and who will quickly find themselves dismayed by the vocabulary.

on the young Earth: there must also have been a process to ensure they were all of the same handedness or 'chirality'.

The first person to tackle this explicitly was a physicist named Frederick Frank.* He had done crucial intelligence work during the Second World War, helping to identify covert radar stations and to alert the British government to the Nazis' secret rocket weapons. He spent most of the rest of his career studying crystals.[8]

In 1953 – the year the Miller experiment and the structure of DNA were published – Frank set out his solution to the handedness problem. In his paper, he made it clear that he regarded the question as trivial.[9] While others thought the 'asymmetric synthesis' of biological molecules was a challenge, Frank wrote, 'I have long supposed that this was no problem.'

His solution rested on the fact that all living things can reproduce themselves. Likewise, Frank suggested, some chemicals can reproduce themselves by catalysing their own formation: we have seen several such autocatalysts. Frank supposed that the first biological molecule, once formed, was such a self-replicator. But he also suggested that this self-replicator might catalyse the formation of new copies of the same handedness, and inhibit the formation of the opposite handedness. Frank ended with the encouraging suggestion that 'a laboratory demonstration may not be impossible'. It would be forty-two years before this challenge was met.

* It is almost disappointing to learn that his middle name was Charles and not Fraser or Frodo.

The person who finally succeeded was Japanese chemist Kenso Soai. With three colleagues he studied a chiral chemical called an alkanol, which was built around a hexagonal ring of atoms.* At the start of the experiment, there was a slight excess of one version – a situation that could easily arise by chance. Then the team mixed in two additional chemicals to get things going. The result was that the two versions of the alkanol began copying themselves. Over several rounds of reactions, the one that slightly predominated at the start completely took over, stopping the other from copying itself and eventually virtually obliterating it.[10] This process is now called the Soai reaction.

Quite how this worked remained a mystery for six years, until an American chemical engineer named Donna Blackmond took a closer look. She had made her name with work on catalytic converters, the devices used in motor vehicles to cut pollutant emissions. However, in the 1990s the biotechnology giant Merck asked her to help them systematically understand the ways organic chemicals behaved. This took her into organic chemistry, and eventually to a fascination with the handedness problem (or chirality, as chemists call it).† The common thread was her expertise in closely tracking the progress of chemical reactions. Since 2001 Blackmond has been involved in a string of seminal studies of the chirality problem, often working with other groups.

In the first instance, Blackmond teamed up with John Brown at the University of Oxford. Their team showed that two alkanols

* The alkanol in question is called 2-methyl-1-(5-pyrimidyl)propan-1-ol, but we'll stick with 'alkanol'.

† Blackmond says her career has taken her 'from auto catalysis to autocatalysis'.

of the same handedness could pair up, and it was this double molecule that drove the reaction.[11] Pairs of alkanols of opposite handedness did not have this effect, so such molecules simply sat there.

The Soai reaction was a crucial proof of principle, showing that chemicals of virtually all the same handedness could form naturally from mixtures of both versions, given only a small starting imbalance. Such an imbalance is not implausible: if nothing else, if there is an odd number of molecules, the two versions cannot be perfectly balanced.

However, there was a problem: the alkanol Soai used was unusual, and the reaction was specific to it. The alkanol also had little connection to biological molecules like amino acids or nucleotides. At first, researchers hoped a similar process would be found that worked for these chemicals. But in over twenty years nobody has found a way to make one version of an amino acid or nucleotide catalyse its own formation, while inhibiting the formation of the other version – at least, not all the way. There may not be a Soai reaction for biological molecules.

A second issue with the Soai reaction was the need for a small excess of one handedness to begin with. The reaction was superb at amplifying this initial imbalance, but could not generate it. It may be that the chirality problem must be solved in two steps: first create a small imbalance by one process, then amplify it by a second process.

Since the 1960s, some researchers have tried to achieve the initial imbalance by way of fundamental physics. The suggestion was that nature might have a subtle preference for left-handed or right-handed versions of chiral molecules, as a result of subtle differences in the nuclei of their atoms.

The idea came about after physicists asked themselves what things would be like in a universe that was the mirror image of ours, in which left was right and right was left – as in Lewis Carroll's *Through the Looking-Glass*. Would everything be the same, albeit swapped around, or would the mirroring inevitably cause other changes? Most thought that everything would be the same, a concept dubbed 'parity conservation', but in 1956 two theoretical physicists named Tsung-Dao Lee and Chen-Ning Yang argued otherwise.[12] The following year, at their suggestion, experimental physicist Chien-Shiung Wu showed that they were right: parity was not always conserved, so mirror worlds would be different.[13] She found that the weak nuclear force, one of two forces that control atomic nuclei, does not obey the parity rule. Lee and Yang were awarded the Nobel Prize in Physics that year.

Almost a decade later, Japanese physicist Yukio Yamagata suggested that this parity violation within the nuclei of atoms would mean that one version of a chiral molecule would have slightly more or less energy than the other.[14] The difference would be tiny, but it would make one more likely to form. This 'parity-violating energy difference' has garnered a lot of attention over the decades, but the effect seems to be too small, by many orders of magnitude. In 1985, Dilip Kondepudi and a colleague suggested that the difference might be amplified by other processes, so that a 50:50 mixture of left- and right-handed molecules would end up all the same handedness within 15,000 years.[15] But this turned out to rely on a lot of assumptions, all of which seem far too optimistic. On the current evidence, parity violation does not explain the initial imbalance.[16]

Instead, other physical processes could be at work. In Pasteur's experiment, the two versions of tartrate could be identified because they formed different-shaped crystals. This means the molecules had segregated: left-handed molecules preferred to crystallise with other left-handed molecules, and right-handed ones also crystallised with their fellows. This is a ready-made way to separate the two kinds of molecule. Simply evaporate the water in which they are dissolved, or add more than can be dissolved, and the molecules will sort themselves into crystals.

A dramatic example of this process was demonstrated in 1990 by Kondepudi and his colleagues.[17] They studied sodium chlorate, which behaves in a curious way: it is not chiral when dissolved in water, but it forms two kinds of chiral crystal much like tartrate. When it crystallises out of still water, it forms equal amounts of left- and right-handed crystals. However, if the water was stirred very fast, over 99 per cent of the crystals were the same handedness. The first crystal to form, dubbed the 'Eve crystal', determined the handedness of those that followed.[18] The trick was that the rapidly spinning stirring bar hit the Eve crystal and shattered it. This created lots of crystals of the same handedness, each of which acted as a seed for more crystallisation. This happened before a second Eve crystal (possibly the other handedness) could form.

Fifteen years later, Spanish chemist Cristobal Viedma came up with a different trick for persuading sodium chlorate molecules to adopt the same handedness – one that so surprised his fellow chemists, it took him a year to get it published.[19] Viedma started by dissolving sodium chlorate in water until no more could dissolve. Crystals formed, half left-handed and half right-handed. The whole setup was at equilibrium: while molecules were constantly detaching from the crystals and reattaching, the crystals stayed about the same size and the same mix of handednesses.

However, this changed when Viedma added glass beads and began stirring the mixture. The glass beads crashed into the crystals, breaking them into smaller pieces – some of which were small enough to dissolve. Suddenly there was too much sodium chlorate dissolved, so the molecules started adding themselves to the remaining crystals. But they did not do so at random: the molecules preferred to latch onto larger crystals. If by chance there were slightly more large crystals of one handedness than the other, those crystals grew faster and swept up ever more molecules – so a tiny imbalance quickly ballooned into a big one.

These were startlingly beautiful experiments, but of course sodium chlorate has nothing to do with life. However, in 2008 Blackmond teamed up with a Dutch group to show that the same process could work for a chemical derived from an amino acid.[20] She and Viedma began working together that same year, demonstrating the same thing for an actual amino acid.[21] One of the essential biological molecules had been persuaded to convert its right-handed molecules into left-handed ones.

This is all very well, but only 10 per cent of chiral molecules form pure crystals like this if both versions are present. The remainder form mixed crystals with equal quantities of left- and right-handed molecules. What about those chemicals?

In 1969, Harold Morowitz offered a possible solution.*

* We have previously met Morowitz supporting the compartmentalisation-first hypothesis (chapter 9) and theorising about the origins of metabolism (chapter 10).

Morowitz pointed out that the different kinds of crystals do not dissolve equally readily in water.[22] Suppose that crystals containing only left-handed molecules are more likely to dissolve in water than crystals containing both left- and right-handed molecules. In this case, the left-handed molecules will accumulate in the solution. Blackmond's team showed in 2006 that this could work for amino acids.[23] Impressively, dissolved serine became 99 per cent left-handed.

These discoveries came in a rush in the early twenty-first century. After decades of groping in the dark, suddenly chemists interested in the chirality problem had several viable mechanisms to work with. Science writer Philip Ball drily noted that they were 'spoilt for choice'.[24]

What's more, the ideas have kept coming. In the last decade, physicist Ron Naaman has shown that left- and right-handed molecules respond differently to magnetic fields. In 2019 his team sorted three amino acids into their left- and right-handed versions using magnets: one version crystallised on the magnet's north pole, the other on the south.[25] Of course, strong magnets are rare in nature, but many minerals have weak magnetic fields.

The lesson from these studies, Blackmond now argues, is that several mechanisms were in play billions of years ago. Each process nudged the first biological molecules away from a 50:50 mix and towards one handedness or the other.

If nature does have ways of pushing chiral molecules one way or the other, we should be able to see the results even in places that are lifeless. In other words, if chemicals like amino acids have formed anywhere off Earth, they should not be 50:50 left and right.

The first strong evidence for this was published in 1997 by

astrochemist Sandra Pizzarello.[26] She and her colleague John Cronin studied samples of a meteorite that fell to Earth near the small town of Murchison in Australia. The rock contained tiny amounts of an unusual amino acid not found in living organisms, and there was significantly more of the left-handed version. Later studies have found excesses for other amino acids of up to 18 per cent.[27] Clearly, at least one of the known mechanisms is at work in meteorites.

However, it is unclear whether organic molecules floating freely in space are also being pushed towards particular handednesses. Most chemicals known from space are quite simple molecules and are therefore not chiral. In 2016 the first exception was found: astronomers detected a chiral chemical called propylene oxide in clouds of dust many light-years away.[28] We do not yet know whether the molecules are 50:50 left and right: merely finding them was a feat. It may be that imbalances only form when chiral molecules interact with rocks and other substances, in which case imbalances will only be seen on planets and other rocky bodies.

Still, in at least some situations – including those most relevant to life on Earth – chiral molecules do tend to move towards a preferred handedness. But that is not the end of the story.

The challenge now is to integrate the new understanding of chirality into the chemistry for making biological molecules and kick-starting life.[29] The two disciplines have hitherto been almost separate. 'The people who worry about prebiotic chemistry don't worry about chirality, and the people who worry about chirality don't worry about prebiotic chemistry,' says Blackmond.

Since 2011, her group has focused on this question. In

particular, Blackmond has examined what happens when two kinds of biological molecules meet. If one is slightly more left-handed than right-handed (say), can that drive the other away from 50:50 as well? The answer seems to be yes.[30] For instance, an excess of one handedness in sugars can drive amino acids, which are initially 50:50, towards a strong excess. The reverse is also true. This is particularly neat because sugars are a key component of nucleotides, so in effect the building blocks of nucleic acids and of proteins help each other towards the correct handedness.

Others have found similarly encouraging effects. As early as 1997, Pier Luigi Luisi's team found that vesicles made of chiral lipids were more stable if the lipids were all the same handedness.[31] Four years later, M. Reza Ghadiri's team showed that a self-replicating protein could select amino acids of the correct chirality, suggesting even quite simple biological molecules could be choosy about chiral chemicals.[32] Similarly, RNA selectively picked up nucleotides of the correct chirality when it was grown on montmorillonite* clay.[33]

Despite all this progress, it might be objected that nobody has managed to get amino acids and nucleotides to be 100 per cent the correct handedness. The answer may be that it doesn't matter. The first life may not have needed all of its molecules the same.

In 2014, Jonathan Sczepanski and Gerald Joyce created a new kind of RNA-based enzyme, which they described as 'cross-chiral'.[34] There was a version made of right-handed nucleotides, and it strung together bits of left-handed RNA to make a second,

* I realise this is getting repetitive now.

left-handed version – which in turn recreated the right-handed version. That such a thing can even exist is remarkable. The most extreme interpretation is that the first life could have used equal amounts of right- and left-handed RNA, and that the move towards 100 per cent right-handedness only came later. That is probably going too far, as the experiment needs carefully selected molecules to work. But perhaps, if nature was already pushing nucleotides towards right-handedness, life could have met it halfway, coped with the few remaining left-handers, and gradually 'discovered' that pure right-handedness was more efficient.

Over sixty years on, it is starting to look as if Frederick Frank was right all along when he wrote that the chirality puzzle 'was no problem'. Not only are there many mechanisms that cause one version of a chiral molecule to outcompete the other, biological molecules seem to have some ability to cope with mixtures of both versions. Oddly, this resilience may have been more pronounced when life was getting started and its mechanisms were rough and ready. As physicist Paul Davies has pointed out, finely honed machines are more sensitive than crude ones.[35] If you want proof, try running an elderly farm tractor on poor-quality diesel, then try the same thing with a modern Formula 1 car, and see which stalls first. In a similar way, the early organisms may have been slow and error-prone, but that same ham-handedness may also have allowed them to cope with some 'wrong' molecules.

There is clearly more to learn about the handedness problem, but it is no longer insurmountable. However, the crucial point is that the solutions also point the way towards solving the origin-of-life puzzle. There are three key clues. First, no single process solved the chirality problem: several are necessary. Similarly, life's complexity suggests multiple processes were needed to start it.

Second, a clunky solution is better than nothing and will work just fine until something better comes along. If life did not have to be 100 per cent the correct handedness from the start, perhaps it could cope with other glitches too.

Third and most crucially, Blackmond's study of sugars and amino acids reveals that the chirality problem is easier to solve if you have several kinds of biological molecule, because they help each other along. This finding hints that, where an RNA World or Protein World would fall down, an RNA–Protein World might stand tall. A more holistic approach to the origin of life was emerging, and in the last two decades it has come to the fore.

Chapter 13

Return of the Blobs

By the early twenty-first century, the competing hypotheses for the origin of life had fought one another to a stalemate.

Proponents of the RNA World could point to experiments showing that RNA could be copied without enzymes, that it could act as an enzyme, and that it made up the core of one of life's most important components, the ribosome. Those who believed protocells must have come first had evidence that such objects formed quickly, given only a supply of lipids, and they could mimic some of the behaviours of cells. Meanwhile, Michael Russell's supporters were delighted by the discovery of alkaline vents, which meant a key prediction of their hypothesis was correct.

There was also the idea that life began with proteins, but even though amino acids were one of the most common products of prebiotic chemistry experiments, this hypothesis had faded into obscurity.

However, despite their successes, the three leading hypotheses had major problems. RNA World supporters had not shown that nucleotides, the building blocks of RNA, could have formed on the primordial Earth. The protocells made by Pier Luigi Luisi and David Deamer could not do much. And Russell had no evidence that life could have harnessed the natural proton gradients within alkaline vents.

Not that these problems stopped anyone advocating for their pet hypothesis. Quite the reverse: the arguments at conferences had reached an intense pitch, from which they have never really receded. The thing is, people often fight hardest for their ideas when they know deep down they aren't quite right. An outside observer would quickly conclude that the problems with the leading hypotheses were all critical. A new approach was needed, one that didn't try to solve everything with one magic chemical or another, and instead embraced Earth's messiness to build something dynamic enough to be alive. Today the proponents of this new angle can look back on an astonishing run of experimental successes.

The story began in the late 1990s when, despite the squabbling, two members of competing groups found something they agreed on. This small *rapprochement* was started by Pier Luigi Luisi, who had been studying simple lipid vesicles (see chapter 9). Aware of the limits of these protocells, in 1994 Luisi and his colleagues decided to make something more elaborate.[1] They took RNA, the enzyme needed to replicate it and some nucleotides, and placed the lot inside a vesicle. The enzyme worked happily, copying RNA inside the vesicle.

This new kind of protocell, they argued, was 'a step towards the construction of a synthetic cell model': one in which 'the reproduction of the membrane and the replication of the internalized RNA molecules proceed simultaneously'. In other words, they had fused the genetics-first RNA World hypothesis with the compartmentalisation-first vesicle hypothesis. Instead of trying to get one of life's key systems going on its own, they made two at once.

The protocells probably bore no resemblance to anything that existed on the primordial Earth. In particular, the RNA-copying enzyme is as elaborate as most enzymes, and could not have sprung up from scratch. But once again it was a proof of principle. Two seemingly disparate hypotheses had been combined, however crudely.

The person who took this idea and ran with it was Jack Szostak of the Harvard Medical School. Facially, Szostak looks rather like the late actor Charles Hawtrey. Happily, the resemblance ends there: where Hawtrey was a bad-tempered alcoholic, Szostak is mild-mannered and well-liked. He is Canadian, albeit born in London, and loved science since early childhood.[2] His father helped him build a chemistry lab in their basement, while his mother supplied him with 'remarkably dangerous chemicals' from her workplace. On one occasion, the young Szostak vented too much hydrogen gas from an experiment, leading to 'an impressive explosion which resulted in a glass tube being embedded in a wooden ceiling rafter'.

His enthusiasm led him to McGill University in 1968, aged just fifteen. But he struggled to find his direction as a scientist, until in 1980 he encountered biologist Elizabeth Blackburn. She had found that the long strands of DNA in living cells had repetitive bits at their ends. These caps are now called telomeres and are still studied today, as they seem to play a role in ageing. Szostak helped Blackburn show that they protect the rest of the DNA.

Three decades later, Szostak, Blackburn and Carol Greider shared a Nobel Prize for their discoveries.

However, even as the telomere papers were being published in the early 1980s, Szostak's attention was shifting towards the origin of life. It was then that Thomas Cech and Sidney Altman discovered the first RNA enzymes or 'ribozymes', a key piece of evidence that the first life relied heavily on RNA (see chapter 8). Szostak found the ribozyme studies 'really cool', so he focused on the RNA World through the 1990s.[3]

Szostak started going to conferences on the origin of life, where he met Pier Luigi Luisi. At first it was not a meeting of minds. Luisi was a believer in compartmentalisation-first and Szostak an RNA World partisan, so their conversations regularly descended into arguments. But over the years both realised that the other had a point. A cell without a gene is in a sense empty, without the ability to pass on information to its offspring and participate in evolution. Meanwhile, a gene without a cell is naked, and incapable of holding onto the other molecules with which it must co-operate.

The first life, they concluded in 2001, must have had both. It must have been an RNA hosted inside a vesicle.[4] Crucially, both components must have been able to copy themselves. Szostak and Luisi further suggested the two needed to be coupled. The simplest way was for some of the RNA to be a ribozyme that made more lipids for the membrane. This would be 'a sustainable, autonomously replicating system, capable of Darwinian evolution', the pair wrote. Such a cell, while simple, would be 'truly alive'. 'The synthesis of simple living cells' had become 'an imaginable goal'.

The idea that it was necessary to put genetic material in a

protocell, right from the start, was not entirely new. Manfred Eigen said as much in 1971 when describing his 'hypercycle' networks of replicating RNAs and proteins (see chapter 6).[5] Eigen argued that such biological molecules would need 'to escape into a compartment' so they could take advantage of any beneficial mutations they acquired. 'Only those systems which managed to compartmentalize and individualize finally had a chance to survive,' he wrote.

Still, given how entrenched the various camps were by 2001, Szostak and Luisi's new stance was markedly different. Both were abandoning their strict beliefs, instead pushing a hybrid approach in which two of life's vital systems emerged together. Szostak soon resolved to put his money where his mouth was and his team started experimenting with protocells. Within three years, he reported his first major success.

It was an elaborate study, containing several experiments that built on each other. The team began by exploring ways to quickly make protocells out of lipids. As we saw in chapter 9, lipids will spontaneously assemble into droplets called micelles. However, micelles do not have interior space where RNA could be housed, so it was necessary to convert them into vesicles, which do. The conversion was slow, until the team found a catalyst to speed it up: montmorillonite. Finally, the full capabilities of this remarkable clay became clear. Szostak's team found that grains of montmorillonite accelerated the conversion of micelles into vesicles a hundredfold. Vesicles became visible within one minute, often with a grain of montmorillonite trapped inside.

This latter point was crucial. We saw in chapter 8 that montmorillonite helps RNA molecules form and extend themselves: they layer themselves onto the surface and grow there. So by creating vesicles that contained grains of montmorillonite, Szostak's team had created the ideal home for RNA. In fact, when they added RNA to the montmorillonite grains, then used the grains to drive the formation of vesicles, each new vesicle contained a grain of montmorillonite covered with RNA. Crucially, the RNA did not leak out.

This was remarkably elegant. A single mineral had helped create elaborate protocells that contained nucleic acids, using only a handful of chemicals.

In further experiments, the team found that the protocells could grow by pulling in additional lipids from their surroundings. This process was sensitive and only really worked if extra micelles were added slowly. But it did work, just as Luisi had found in the 1990s.

The team also found a way to persuade the protocells to divide, forming 'daughter' protocells – rather like how living cells reproduce by splitting in two. They squeezed large vesicles through tiny holes in sheets, forcing them into long, thin sausage shapes. These were unstable and quickly shattered into many small vesicles. This was not the same as normal cell division, in which one cell splits into two, but crucially the vesicles did not lose much of their precious RNA. In the final experiment, the team put the protocells through repeated cycles of growth and division, like a culture of bacterial cells growing in a lab.

Szostak's protocells did not contain any proteins, enzymes or other biological machinery. Nevertheless, they were astonishingly lifelike. 'These experiments constitute a proof-of-principle demonstration that vesicle growth and division can result from simple physico-chemical forces, without any complex biochemical machinery,' the team concluded. If anything, this was underselling it. As we saw in chapters 4 and 6, modern living cells are complex, with thousands of components all working in tandem. But Szostak's protocells mimic many of living cells' fundamental abilities, despite only containing a few chemicals. The physicist Niels Bohr is often quoted as saying that anybody who isn't shocked by quantum mechanics hasn't understood it. The same might be said of Szostak's experiment: it is shocking how closely the protocells mimicked life, considering how simple they were.

The study was published in 2003, fifty years after Miller's seminal experiment showing that biological molecules could form naturally.[6] After several decades in which investigations of the origin of life had repeatedly inched forwards only to become mired in arguments, it represented a seismic step forward – both as a practical experiment and on a conceptual level.

Over the next decade, the team showed that their protocells were even more versatile than at first thought.[7] Just one year later, they showed how closely the RNA was working with its lipid housing. Szostak and Luisi had suggested linking the two by introducing an RNA enzyme that would make new lipids. But now Szostak's team found a simpler link.

When a vesicle housed lots of RNA, the RNA exerted pressure on the membrane, stretching it like heavy shopping in a plastic carrier bag. Szostak's team found that these overstretched protocells could take lipids from neighbours that lacked RNA. In effect, the protocells competed for their lipid building blocks, and those with lots of RNA won. The fundamental physics of membranes pushed those with RNA to grow and those without to shrink. This simple competition, they wrote, 'could have played an important role in the emergence of Darwinian evolution'.[8] For instance, protocells whose RNA copied itself faster would necessarily grow faster.

The protocells also proved resilient: they could withstand being cooled to 0°C or heated to 100°C.[9] This implied that they could have survived in environments like geothermal ponds or hydrothermal vents. Furthermore, heating them unleashed their capabilities. When they were hot, small molecules like nucleotides could enter them freely, which did not happen at low temperatures. This meant the protocells could 'feed' by taking in new materials when they were hot.

However, there was a problem. How could the protocells divide and thus reproduce?[10] In the original experiment, this was achieved by forcing the protocells through tiny holes to disrupt their shape, but this was highly artificial and unlikely to have happened billions of years ago. Besides, squeezing the protocells through holes caused them to lose some RNA. There had to be a better way.

This problem was cracked twice by the same student: Ting Zhu. In 2009 he and Szostak made protocells with several layers of membrane, like the layers of an onion.[11] When these protocells were fed lipids, they grew into long strands. These were delicate, so a gentle water current was enough to make them divide into dozens of new protocells, without spilling their contents. Three years later Zhu found a second method.[12] He fed the sausage-shaped vesicles with simple chemicals, then shone a light on them. This triggered chemical reactions, causing the vesicles to divide. It was now possible to imagine the protocells growing and reproducing largely on their own.[13]

However, persuading the protocells' RNA to copy itself was a bigger problem. This had to be done without a complex enzyme. Somehow, nucleotides had to line up along an existing RNA and link up to form a new strand. Researchers like Orgel had struggled to achieve this 'non-enzymatic replication' since the 1980s. Now Szostak needed to make it happen in a protocell.

By 2012 he was tackling it in earnest with his student Katarzyna 'Kate' Adamala. There were many ways it could go wrong: for instance, nucleotides sometimes attached themselves to the RNA the wrong way round.[14] Nevertheless, the following year the pair achieved their first success.[15]

They knew RNA copied itself faster if it was mixed with magnesium ions, which is plausible: magnesium is common. Unfortunately, magnesium also destroyed the protocells' lipid membranes. Adamala and Szostak solved this by adding citrate, a chemical almost identical to the citric acid in lemons. Citrate is found in all living things, and it bonded to the magnesium. Now the magnesium accelerated the copying of RNA, without destroying the protocells. The combination of magnesium and citrate allowed the RNA in the protocells to start copying itself.

It has since emerged that iron can be even better than magnesium at speeding up RNA self-copying. Szostak demonstrated this in 2018, working with several colleagues including Adamala – now running her own laboratory.[16] This was excellent, because iron was probably common in the oceans when Earth was young. It is rarer now, because oxygen in the air has reacted with it, but at first there was no oxygen.

It is worth noting that the RNAs in the protocells are not genes. Unlike the nucleic acids in modern organisms, they do not code for anything. If the sequence of bases matters at all, it's because it endows the RNA with some useful ability. Szostak's protocells have nucleic acids, but not genes. However, this is arguably a feature rather than a bug, as having the RNA in place means it could later be co-opted to form genes. Indeed, prebiotic chemist John Sutherland (see chapter 14) says the ignition of non-enzymatic RNA replication in Szostak's protocells is one of the biggest advances in the study of life's origin in years.[17] 'He's made a huge amount of progress,' Sutherland says. 'I think that's dramatic.'[18]

However, not everyone is convinced. A recurring criticism has been that the protocells, ingenious though they are, are implausible because they rely on pure laboratory chemicals – whereas all that was available on the young Earth was a mess of hundreds of chemicals. Szostak's team has responded by showing that mixtures of lipids can still form protocells, and can in some ways make it easier.[19]

It is also true that the protocells do not have anything resembling a metabolism. They can grow by taking in new lipids from their surroundings, but they cannot digest anything or make new chemicals of their own.

In the last decade Szostak has begun tackling this, principally by adding catalysts and simple proteins to the protocells.[20] For instance, in 2013 he and Adamala added a protein containing just two amino acids.[21] This mini-protein was a catalyst that created a second one – and this second protein promptly attached itself to the protocell's membrane and helped it grow. In this way, protocells with simple proteins could grow faster than those without: the more complex protocells had the advantage, suggesting the chemistry could drive them to become more intricate. Similarly tiny proteins helped RNA to attach itself to the membrane, where it was more likely to undergo useful reactions like copying.[22]

Still, these are baby steps and a true metabolism seems a long way off. Even the simple Wood–Ljungdahl pathway, which Bill Martin has proposed as the first metabolic process, is significantly more intricate. It is not yet clear if such a process could have arisen in the protocells.

However, while it is true that Szostak hasn't yet managed to make life from scratch, neither has anybody else, and it would be foolish to ignore the conceptual progress underlying his work. By suggesting that a simple protocell could have a membrane, nucleic acids (albeit not yet working as genes) and maybe even a metabolism, Szostak is pushing for the fusion of three components that have long been treated as separate.

He is not the first to suggest such a thing – and he acknowledges this.[23] The idea can be traced to the 1970s and the work of a Hungarian theoretical biologist: Tibor Gánti. His 'chemoton' model describes a minimal form of life strikingly similar to what Szostak now has in mind. Gánti's ideas languished in obscurity for decades, before recognition finally came in the late 1990s.

Born in 1933, Gánti was fascinated by nature from a young age, in particular with the question of the essence of life: what distinguishes living from non-living matter?[24] Concluding that this was a chemical question, he studied chemical engineering. From 1958 to 1974 he was an industrial biochemist, but also studied microbiology and somehow wrote what was at the time the only textbook of molecular biology available in Hungary.

In 1971 Gánti published *The Principles of Life*, which contained the earliest description of the chemoton model.[25] Unfortunately, it was only published in Hungarian. Even in Hungary his ideas were received with 'complete lack of interest, incomprehension, ridicule and malevolence', according to his student and supporter Eörs Szathmáry.[26] The model was in any case incomplete, and Gánti revised it over several years.[27] A second book followed in 1979 and was republished in English, but again hardly anybody noticed.[28] It was only in 1995, when Szathmáry included Gánti's work in a much-read paper outlining evolutionary history, that the chemoton model became better known.[29] Finally, in 2003 *The Principles of Life* was republished in English and the Gánti revival began in earnest.

The underlying thought behind the chemoton model is that genes on their own, a metabolism on its own, or a membrane-based protocell on its own cannot achieve much. The essence of life is the interaction of all three.[30] While most researchers interested in the origin of life were dividing life up into its subsystems, in the hope that one of them might be enough to get life started, Gánti instead tried to imagine the simplest possible organism that had all three.* This, he argued, was the simplest thing that could be called alive.

In Gánti's model, the metabolic system is a cycle of chemical reactions that is self-sustaining. This repeating process gives rise to the key components for the other two systems: the genes and the membrane. Meanwhile, the genes are long molecules – perhaps RNA – carrying sequence information. They copy themselves by stringing together smaller molecules, releasing by-products that are used in the membrane. This last point may seem unimportant, but it is crucial because it represents the

* Joan Oró and Antonio Lazcano made a similar suggestion in 1984, apparently in ignorance of Gánti's ideas.

genes taking control of the other systems. The faster the genes copy themselves, the faster the membrane builds up – until the protocell is ready to reproduce by dividing in two.

If a chemoton is the simplest possible form of life, Szostak is currently about two-thirds of the way to creating one in his lab. He has genes copying themselves within a membrane-based protocell, and has found ways to link that copying to the growth and division of the protocell. The metabolism is the missing element, as it is the least developed part of his protocells, which by Gánti's definition are not yet truly alive.

What would it take to incorporate a metabolism into Szostak's protocells?[31] It is probably asking too much that such simple structures make all their component chemicals from basic raw materials, but perhaps they could make some of them – or start making some other critical molecule like a simple protein. Perhaps the RNA acquired the ability to absorb energy from sunlight, then used that energy to make more copies of itself.[32] Theoretical simulations suggest that such a 'metabolic replicator' would outcompete non-metabolic RNA.[33] Alternatively, there is evidence that sets of RNA molecules can make their own ribozymes by breaking up useless strands of RNA and reassembling them.[34] It is also conceivable that the metabolic reactions highlighted by Wächtershäuser or Martin might work in the protocells. One idea being examined is that the first cells stored energy in simple chains of phosphates, rather than the ATP favoured by modern organisms (see chapter 11).

Or perhaps this is overcomplicating things. Metabolism boils down to the ability to control chemical reactions, so certain processes happen and others don't. That means having catalysts to speed up the desired reactions. For modern organisms that means enzymes. But many enzymes have at their cores something simple: single atoms or clusters of particular metals. One common arrangement is a clump of iron and sulphur atoms. In 2017 a team led by Claudia Bonfio, and including Szostak, showed that such iron–sulphur clusters could attach themselves to simple proteins – inside protocells.[35] The combination of iron and sulphur recalls Günter Wächtershäuser's Iron–Sulphur World from chapter 10.

These avenues all seem promising, but they might prove tricky in practice. For that reason, it would be foolish to predict how soon we will build a complete chemoton. Still, it does not seem like an impossible dream. Such a minimal cell would be the most plausible starting point for life on Earth ever developed.

One might ask why, if such a minimal organism can exist, we have never found one. Even the simplest bacteria have hundreds of genes and are orders of magnitude more complicated than Szostak's protocells. One possible answer is that complex organisms are more versatile and resilient. In that case, chemoton protocells survived while their only competition was other chemoton protocells, but they have long since been ruthlessly outcompeted by more elaborate organisms. Indeed, several of Szostak's experiments show more elaborate protocells outcompeting their simpler cousins.

It is also worth remembering that our knowledge of the microbial world is still woefully incomplete. Perhaps the most pertinent illustration of this is the discovery of giant viruses. As the name suggests, they are much larger than normal viruses: some are the size of a bacterial cell. They were described in 2003: the first, now called *Mimivirus*, had been found in 1992 but mistaken for a bacterium.[36] Unlike most viruses, giant viruses carry quite a few genes, including some that code for the machinery that reads genes and makes proteins.[37] They still need to invade a living cell to reproduce, but otherwise they are a halfway house between 'traditional' viruses and cells. Perhaps they are cells that have adopted a parasitic lifestyle; maybe they were once simple viruses that have evolved greater complexity; or perhaps they are something unique.[38] It is too early to decide, but one thing is clear: giant viruses blur the line between living and non-living still further. In some ways they are more complicated than a chemoton, but their inability to reproduce independently nevertheless marks them as non-living. Their existence should open our minds to simpler forms of life.

The experiments described in this chapter bring us much closer to the day when we can make a simple living organism from scratch. Gánti's ideas and Szostak's experimental ingenuity both point the way towards a self-assembling minimal organism. Their success is arguably the strongest possible rebuttal to earlier ideas like the RNA World, and suggest that life can only form when all its components are available.

But, of course, Szostak's experiments do rely on a ready supply of chemicals like lipids and nucleotides. Where, his critics ask, do they come from? And so the story comes full circle, back to the question Stanley Miller tried to answer in the 1950s: how did the chemicals of life form? This is the subject of the final chapter. We will see that the same principle of 'everything at once', which has helped create lifelike protocells, also makes it easier for life's building blocks to come into being.

Chapter 14

Just Messy Enough

If life began with a chemoton-like protocell, it had to have all the necessary chemicals: nucleic acids, lipids and probably proteins too. The person who has arguably done most to show how this life-giving cocktail might have come to be is chemist John Sutherland.

His interest in life's beginning dates to his childhood in the 1960s. 'I was always interested in where we came from,' Sutherland says. He could not study the origins of the universe, which needed advanced maths, so when he went to the University of Oxford in 1980 it was to study chemistry.

Like Stanley Miller before him, his life was changed by a lecture: one given in the mid-1980s by chemist Albert Eschenmoser, who later created an artificial nucleic acid (see chapter 8). Eschenmoser asked why some biological molecules are difficult to make. He argued that this difficulty was not obviously measurable – say, by counting the number of atoms in a molecule – because some structures self-assemble given the right ingredients and others do not. 'However complex chemicals look, if they self-assemble then the complexity is in the eye of the beholder,' says Sutherland. 'It was a "road to Damascus" moment. It was obvious that this is something you ought to apply to RNA.'

At the time, the RNA World hypothesis was ascendant, but there was a snag. It was proving tricky to make nucleotides, the building blocks of RNA. This suggested they could not have formed naturally on the Earth, in which case the RNA World was a non-starter.

Sutherland thought otherwise. He became convinced that the apparent complexity of RNA nucleotides was an illusion and there was a way to make them. But it took him two decades to find it. 'We gradually got some money to do some work: never that much, but God bless the research councils, they gave us some,' he says.

The eventual solution required lateral thinking. Biochemists thought of a nucleotide as three smaller bits joined together: a base, a sugar and a phosphate. These subdivisions are obvious if you look at a diagram of the molecule. It followed that the way to make a nucleotide was to build these three components and link them up. It was neat and logical, and it didn't work. The problem was that the sugar and base wouldn't join up. The molecules were the wrong shape, so like two mismatched jigsaw pieces they would not fit together.

Sutherland, with Matthew Powner and Béatrice Gerland, found another way. Instead of making sugars, bases and phosphates, they started with five simple substances. One was cyanamide: the same cyanide-like chemical that Joan Oró used to make biological molecules (see chapter 3), and which David Deamer used to make lipids (see chapter 9). When the team put these five chemicals through certain reactions, nucleotides formed. But at no point did the team make bases or sugars. If the traditional approach was like assembling a skeleton by first building the limbs, ribcage and head, Sutherland's method entailed building half of each limb, a ribcage with bits of limb attached, and a head, then stitching them together.

The experiment was published in May 2009 to glowing coverage.[1] Szostak called it a 'synthetic *tour de force*' that 'revive[d] the prospects of the "RNA first" model'.[2] Many others saw it in a similar light: a vindication of the RNA World.

However, Sutherland did not interpret it that way. He accepted many of the arguments that RNA preceded DNA as the genetic material, but was always more inclined towards the 'soft' RNA World hypothesis, in which RNA did not perform all the functions of life on its own – an idea that struck him as unrealistic. It would be better, he reasoned, to show that RNA emerged from the same chemicals that gave rise to amino acids (and thus proteins) and lipids. 'We started with an emphasis on RNA, but with a mind that was open, and we hoped that we could get everything at the same time,' he says.

It would be hard to exaggerate how much this went against conventional thinking, particularly in the 1990s when Sutherland ran his first, abortive experiments. RNA is a complex molecule, the thinking went. So are proteins and lipids. Surely they would not all self-assemble from the same chemicals? No, they must have formed separately, from distinct building blocks and under different conditions, and then found their way to each other – perhaps on ocean currents. This assumption was so ingrained that it was never explicitly stated, let alone examined. The alternative simply seemed unimaginable.

This idea goes back to the chicken-and-egg paradox from

chapter 6. In modern life, DNA makes RNA using a protein enzyme, and the RNA is then used to make protein – by way of a ribosome containing RNA and protein. Some of those new proteins are then used to maintain and copy the DNA, closing the circle. In other words, RNA and protein are interdependent. The RNA World hypothesis was supposed to cut this Gordian knot by showing that RNA could perform the whole cycle on its own. But for Sutherland, this created more problems than it solved. 'You think you're simplifying it, but you're not,' he says. Better, he thought, to make RNA and proteins from the same melting pot, so they could co-operate straight away.

The key to this lay in more complex mixtures of chemicals than had hitherto been used in prebiotic chemistry – but not *too* complex. Put too many carbon-based chemicals in the same pot and the result was a sludgy tar that was no use to anyone. Meanwhile, too few chemicals would mean only a handful of reactions and nothing very impressive being made. What Sutherland needed was a Goldilocks chemistry: neither too simple nor too complicated.

His RNA nucleotide synthesis was a prime example of this. The last component that was bolted onto the fledgling nucleotide was the phosphate, but Sutherland found that the phosphate had to be included in the reaction mixture from the start, because it stopped certain undesirable reactions. Putting the phosphate in so early would strike many chemists as unduly messy, but it actually helped.

This approach of harnessing complicated mixtures has been dubbed 'systems chemistry' by Günter von Kiedrowski, who we met in chapter 8 when he made self-replicating sets of molecules.[3] The idea is not dissimilar to the reaction networks envisioned by

Stuart Kauffman, who saw that many chemicals together will behave in surprising and intricate ways, whereas a bare handful will do little. But the end result must not simply be a mess of hundreds of chemicals. The aim is to find a combination that makes lots of the things you want, and little else.

In fact, the evidence that nucleotides, amino acids and lipids could all form together has been accumulating, underappreciated, for decades. One line of evidence comes from meteorites that have fallen to Earth, like the Murchison meteorite that landed in Australia in 1969. The Murchison meteorite is a special type called a carbonaceous chondrite. These are rich in carbon chemicals, including many of those associated with life. As early as 1985, David Deamer showed that there were lipid-like molecules in the Murchison meteorite, which could form membranes and vesicles.[4] Many amino acids have also been found,[5] and astrobiologist Zita Martins has identified one of the bases from RNA in the Murchison meteorite.[6] The biological molecules are not plentiful, but the fact they are all present in the same extraterrestrial rocks hints that they can form together, given the right conditions.

What process might have created all these chemicals at once? The Italian biochemist Ernesto Di Mauro has spent much of his career on this question. In the 1990s, he began studying a chemical called formamide.[7] This is one of several simple chemicals related to cyanide, all of which have proved useful for making biological molecules.[8] A molecule of formamide contains just six atoms: a carbon, an oxygen, a nitrogen and three hydrogens. It is similar to the cyanamide used by Sutherland, and to formaldehyde; a four-atom molecule that Oró and Wächtershäuser both used.

Crucially, all these chemicals are commonplace in the universe. Astronomers have known that there is formamide, cyanamide and formaldehyde in space since the 1970s.[9] They form from even simpler chemicals like water and hydrogen cyanide, which are plentiful in the darkness between the stars.

Formamide's possible significance in the origin of life has been known since the 1960s.[10] Then in 2001 Di Mauro's team showed it could be transformed into some of the building blocks of RNA. They heated pure formamide to 160°C for forty-eight hours, in the presence of common minerals like limestone, and got adenine and cytosine: two of the bases found in RNA and DNA.[11] Of course, this was not as impressive as getting entire nucleotides, as Sutherland would do eight years later. But at the time it was a significant step.

'Since then, we've started analysing all the synthetic reactions that could take place from formamide,' says Di Mauro. 'We found a lot of them.' Crucially, many minerals can speed up the formamide reactions,[12] including – and at this point this should not surprise anyone – montmorillonite clay.[13] That means formamide can do its thing in many different locales, rather than being dependent on a particular place or mineral.

Furthermore, it does not just make the bases from nucleic acids. In 2011, Di Mauro's group exposed formamide to rock samples from the Murchison meteorite.[14] This time, as well as getting bases, they obtained amino acids. Clearly, Sutherland was on the right track: the building blocks of both proteins and nucleic acids formed from the same simple chemical. Four years later, Di Mauro's team repeated this experiment and added beams of high-energy protons. This extra kick of energy led to the formation of near-complete nucleotides: only the phosphate was missing, and the hard part of getting the sugar and base to link up had been achieved.[15] The entire surface of the young Earth was 'a factory of organic compounds', Di Mauro says.[16]

The obvious point of difference between Sutherland and Di Mauro is their choice of starting material: cyanamide versus formamide. However, while the difference is not trivial, nor is it earth-shattering, because the two chemicals are so closely related: both are derived from hydrogen cyanide. 'Our stories are complementary,' says Di Mauro. Formamide is known to be common but gives lower yields of the key chemicals, whereas cyanamide is rarer but produces a lot. But the results are still bases, sugars and amino acids. 'At the end the chemistries are always the same,' says Di Mauro.

What of Sutherland? After the 2009 paper he moved to the Laboratory of Molecular Biology in Cambridge, which does not require academics to publish new findings at a punishing rate, instead allowing them to take risks on long-shot experiments. Sutherland seized the opportunity to hunt for his Goldilocks chemistry. In 2012 he and his colleague Dougal Ritson made two simple sugars called glycolaldehyde and glyceraldehyde, which he needed to make the RNA nucleotides. The pair found the sugars formed from hydrogen cyanide irradiated with ultraviolet light.[17] This was particularly beautiful, because hydrogen cyanide was the source material for the cyanamide on which his reaction also depended.

Three years later, Sutherland's team produced arguably their most impressive experiment to date.[18] The starting point was again hydrogen cyanide, this time alongside a sulphur-based chemical and a mineral that acted as a catalyst, all bathed in ultraviolet light.[19] Depending on exactly which combination was used, the hydrogen cyanide transformed into an array of chemicals. These included the precursors to nucleotides from the 2009 synthesis, precursors of amino acids and even precursors of lipids. The same simple chemistry could be steered towards proteins, nucleic acids or the lipids needed to make membranes: all the key components of living cells.

'RNA and proteins are inseparable in modern biology, and it looks from the chemistry that they were born together,' says Sutherland. In this he was echoing Carl Woese, who foresaw the centrality of the link between nucleic acids and proteins in 1967.[20] Woese pointed out that a gene that isn't used to make a protein isn't really a gene: it's just a strand of nucleic acid carrying meaningless information. Therefore, it did not matter which came first. The key thing was the relationship.

Many remain unconvinced by the idea that all the chemicals of life can form together. For some, Sutherland's synthesis requires too many steps to be plausible. Alkaline vent proponent Mike Russell is less measured: 'Sutherland is just a lunatic,' he told me. 'He's doing all these fantastic bits of chemistry, but none of them mean shit.'

However, others have started exploring similar everything-first chemistries. For example, Leroy 'Lee' Cronin has found a way to make nucleotides from sugars, phosphates and bases – the very reaction Sutherland sidestepped. Cronin made it work in 2019 by heating the three chemicals in the presence of amino acids.[21]

Meanwhile, Loren Dean Williams and Nicholas Hud have reassessed one of the key pieces of evidence for the RNA World. They say it actually points to a mix of RNA and protein right from the start. The evidence in question comes from ribosomes: the molecular machines that read genes and, based on the instructions therein, make proteins. In chapter 8 we saw that the ribosome's core is made of RNA. When this was uncovered in 2000, it was interpreted as evidence for the RNA World. But Williams and Hud point out that the ribosome's function is to link amino acids into proteins, implying RNA coexisted with proteins. By comparing ribosomes from many species, they have reconstructed how the ribosome evolved.[22] It seems even the earliest RNA-based version could string amino acids together – albeit in a random order. The resulting simple proteins would have resembled those made by Sidney Fox (see chapter 7).

Williams argues that RNA and proteins have always worked together. He compares this co-operation to the way many animals work together, like the ants that guard aphids from predators in exchange for sugary drinks. Ecologists call this co-operative behaviour 'mutualism'. For Williams 'little in biology makes sense except in light of mutualism'.[23]

There are more advantages to this way of thinking. If we accept that the chemicals of life emerged together, we can solve a long-standing conundrum.

As we saw in chapter 6, researchers have struggled to decide whether genes or metabolism came first. But once you accept that RNA and protein worked together from the start, this distinction is revealed as meaningless. All the genes-first models, like the RNA World, are actually models in which nucleic acids run a metabolism – or, by copying themselves, *are* a metabolism – as well as carrying genetic information. Meanwhile, it has not yet proved possible to create a self-sustaining metabolism without a nucleic acid.

Indeed, the distinction between genetics-first and metabolism-first collapses altogether once you consider it on a deeper level. Genetics-first emphasises the importance of storing information on a molecule and passing it on. Meanwhile, metabolism-first emphasises disorder and entropy: in particular, the need to harness an energy supply so the organism doesn't fall apart. These may seem distinct, but when you get right down to it, information and entropy are two sides of the same coin. Disorder is what happens when information is disrupted. Life needs information, but that is just another way of saying life is a complicated structure – whether an RNA molecule or a metabolic cycle – that has to be maintained.

The fact that entropy is the opposite of information was revealed by a thought experiment called 'Maxwell's demon'.[24] The idea is that a microscopic person was stationed at the door between two chambers, which contained a mix of two kinds of gas.* By opening and closing the door at precisely timed moments to let individual gas particles through, the demon could sort the two gases so that each chamber only held one kind. This reduced the entropy of the two chambers, without using any energy – something the Second Law of Thermodynamics says is impossible. The solution to this apparent paradox was that, for the demon to do its job, it had to know which particles were which. This meant it had to conduct measurements, remember them, and then delete those memories to make way for new ones. This activity would use energy and thus increase the overall entropy. The lesson for the origin of life is that the amount of

* A similar imaginary setup, without the demon, was discussed in chapter 10.

entropy in the system is completely tied up with the amount of information, and it is not useful to talk about one but not the other.

The biochemist Addy Pross has made a similar point. Pross says that the genes-first/metabolism-first row boils down to 'what was the first thing that could copy itself?'[25] If it was a single molecule, it was probably a chain molecule like RNA, and superficially looked like a gene. If it was a set of molecules that reproduced as a whole, it resembled a metabolic network. But really the distinction is between a single self-replicating molecule and a set of molecules that self-replicated together. Pross points to Gerald Joyce's experiments in which two RNAs self-replicate as a pair, each making the other. This could equally well be interpreted as genes copying themselves, or as a simple metabolic cycle, and we only confuse ourselves by trying to label it one or the other.

So far so good: it now looks plausible that genetics and metabolism, in the form of RNA and protein, can emerge together. But what about the third component of a chemoton-like protocell: the actual cell? It turns out the idea of everything-first has an unexpected supporter: David Deamer.

As we saw in chapter 9, Deamer's focus in the 1980s and 1990s was on making lipids and assembling them into vesicles: simple protocells that could house things like RNA. In other words, he subscribed to the compartmentalisation-first hypothesis. However, nowadays he does not think lipids formed before nucleic acids or proteins. 'Why should we try to think of one of these coming first?' he asks. 'That just complicates the situation, because we know the early Earth must have been a very complex mixture of these things.' The question is how these chemicals 'began to have biological functions'.

Deamer's answer is that membranes and vesicles were the first *structures* to form, simply because lipids assemble into them spontaneously as soon as there are enough in the same place.[26] But once this happened, it made it easier for the other big molecules to form: for nucleotides to link up into RNA and amino acids into proteins. 'Lipid membranes are not just compartments, they are organising principles,' says Deamer. He describes this in his 2019 book *Assembling Life*.[27]

The key, Deamer says, is for the mix of chemicals to experience repeated cycles of wetting and drying. Imagine a small pool on an island. During the heat of the day the water evaporates and the pool dries out, leaving a sticky mess of chemicals clinging to the rock. Later in the afternoon it rains and the pool refills. 'These wet–dry cycles are everywhere,' says Deamer.

The effects of repeatedly wetting and drying the chemicals can be profound. In particular, lipid vesicles become crushed together as the water level falls, and the lipids rearrange themselves into layers of membranes stacked on top of each other. Within these layers, nucleotides and other molecules become trapped. Forced close together by the enclosing lipids, they are more likely to link up.[28]

For instance, Deamer's team has found that nucleotides in such a setup string themselves together into chains resembling RNA.[29] They have also persuaded DNA to copy itself without an enzyme – the same process Jack Szostak worked so hard to achieve with RNA (see chapter 13).[30] 'That went over with a giant thud,' Deamer recalls. 'Nobody believed it.' However, he has since closely analysed what happens in the lipid layers, demonstrating how the nucleotides come together.[31] Furthermore, the bases and sugars in RNA can stabilise groups of lipids, helping protocells to form.[32]

'We really can make stuff happen,' Deamer says. 'We can start with a huge chaotic mix, but between self-assembly and selection, out of that mixture can come fairly specific and interesting organised entities.'

The approaches taken by Sutherland, Di Mauro and Deamer do not precisely map onto each other, but the lesson is the same. Forget the RNA World, the Iron–Sulphur World, the Lipid World and all the other hypotheses that assume a single substance could kick-start life. These models are all doomed to fail. Instead, imagine short chains of RNA and small proteins working together, all bounded by simple lipid membranes. What makes the chemicals of life special is that they are good at co-operating. Omitting one just makes things harder.

The other message from this research is that the formation of life is not as astronomically unlikely as it once seemed. Certainly, a simple cell assembling itself atom by atom is staggeringly implausible. But the chemical processes that have been discovered work readily: lipids spontaneously assemble into vesicles, and RNA copies itself given the right environment. That means we cannot infer the likelihood of a cell forming by counting how many parts it has. The real question is, how specific are the circumstances that unleash the spontaneous processes? We cannot yet put a figure on this, but the processes seem pretty robust.

Still, it may seem unlikely that pure RNA or protein would have formed in sufficient quantities. Surely other, less useful molecules would have turned up and thrown a spanner in the works?

The answer is two-fold. First, processes like Deamer's wet–dry cycles improve the chemicals step by step, so they may have progressed towards purity. And second, a little bit of sloppiness may not have been disastrous. Deamer has found that vesicles made of more than one kind of lipid are more stable.[33] Mixtures containing more chemicals are more likely to display complex, lifelike behaviours.[34] Similarly, Szostak has shown that RNAs can still fold up to form enzymes if some of the nucleotides are, by normal standards, upside-down.[35] He has also made working molecules combining nucleotides from DNA and RNA,[36] and shown that DNA nucleotides can form through similar processes to those Sutherland found.[37] Finally, there is the 1994 discovery by Ronald Breaker and Gerald Joyce that DNA can act as an enzyme – the very ability that made chemists alight on RNA as a precursor.[38]

What is needed, in Szostak's words, is 'a middle ground between the unconstrained and the oversimplified'.[39] Carl Sagan expressed a similar thought in 1963, wondering whether 'the purity of laboratory reagents obscured for us the true sequence of reactions in primitive, and less pure, times?'[40] Evidence for this has now emerged from the study of networks of RNA molecules. We saw in chapter 8 that groups of RNAs can form autocatalytic sets, in which one molecule makes a second, which makes a third, and so on until the first one is recreated and the whole set is copied. In 2019, Ryo Mizuuchi and Niles Lehman* used

* Lehman has made many contributions to biochemistry, but he had been keeping a disturbing secret. In July 2019 he was sentenced to two and a half years in prison for possession of child pornography.

computer simulations to show that such a self-reproducing set was most likely to form if the RNAs were fairly diverse.[41] Too few kinds of RNA meant there probably wasn't an autocatalytic set; too many meant the whole thing went haywire.[42]

In other words, the first life might have had a degree of complexity, provided it was also tolerant of errors. Suppose the first organism contained about fifty kinds of molecule. It seems reasonable to suppose that fifty different molecules could be gathered in the same place – particularly since they all form from the same starter chemicals. Such an organism only becomes implausible if it is unduly fragile: for instance, if removing a single chemical is enough to kill it. But suppose instead that each component is somewhat replaceable. The evidence tells us it is not necessary to have genes of pure RNA: there could have been some DNA, or TNA, or any of the myriad other nucleic acids. Such an organism would be clumsy and slow, a shonky prototype made of a hodge-podge of parts kludged together. It would never survive today: other microorganisms would literally eat it for breakfast. But at the beginning there were no predators.

The first enzymes may have been particularly clunky. As we've seen, RNA can fold up into enzymes called ribozymes. In 2002, John Reader and Gerald Joyce created a ribozyme that only contained two of the four bases.[43] Nevertheless, the ribozyme could join two RNA molecules to make a larger one. The implication is that crude ribozymes could have formed, and worked, even if major components were missing.

There is another way in which the first cells might have coped with their own clunkiness. The clue is in the phrase 'the first *cells*'. Often when people imagine the beginning of life, they envision a single organism existing in lonely isolation. But this seems unlikely for two reasons.

First, if the formation of life was probable – and the evidence in these last two chapters suggests it was – then presumably lots of life formed. Consider Deamer's lipid protocells drying out into layers and then being rewetted, giving rise to hundreds of protocells. The first cell was not alone: it belonged to an instant community.

And second, all life today is intimately bound up with other life. As Harold Morowitz wrote:[44] 'Sustained life under present-day conditions is a property of an ecological system rather than a single organism or species.' Or, to rewrite John Donne, no organism is an island. An individual organism may seem to stand apart, but this is an illusion. Imagine a horse standing alone in a field. Except the horse is not alone: its body is home to millions of microorganisms, many of them essential to its wellbeing. Furthermore, it needs to eat, so it is utterly reliant on plants like grass – and plants need soil, which is made by yet more organisms. Finally, if the horse is to pass on its genes to a new generation and thus participate in evolution, another horse is required. The horse is part of an ecosystem and it cannot survive or reproduce alone. In fact, every organism relies on its neighbours, in a nested set of interweaving cycles that envelop the planet.

It follows that to ask how life began is not simply to ask how one organism formed, but rather to ask how the first ecosystem formed.

The idea that life has always been an ecosystem was explored by Sarah Voytek and Gerald Joyce in a 2009 experiment. They created two RNA enzymes, each of which could copy itself by using certain 'feedstock' chemicals.[45] However, one was better at using a particular feedstock, while the other was better on a second feedstock. If they were only given one feedstock, the faster enzyme thrived while the other died out. But if they were given both, the enzymes coexisted and even evolved to focus on their preferred feedstock.

The enzymes were obeying a law of ecology called the competitive exclusion principle, which says neighbouring species evolve different lifestyles so they don't compete directly. Darwin's finches on the Galapagos Islands are an example. On some islands several finch species coexist, but they all have different diets, reflected in the shapes of their beaks.[46] In a sense, the finches are behaving in the same way neighbouring molecules once did on the young Earth.

This may seem like a last-minute complication, but it makes the problem easier. If there were multiple organisms living side-by-side from the start, then it did not matter if some lacked key systems, because the cells inevitably worked together. There was no conscious intention: simply the fact that the first cells were bad at keeping chemicals inside themselves, so no cell was working in isolation. If a cell could not make a particular nucleotide, probably one of its neighbours could, and that neighbour would be leaky, so some of the nucleotide would escape into the surrounding water, ready to be picked up.

Carl Woese saw this in the 1990s: 'The universal ancestor is not a discrete entity. It is, rather, a diverse community of cells that survives and evolves as a biological unit.'[47] Each cell could only do a few jobs, but between them they got everything done. It is the same idea as Kauffman's autocatalytic sets, but applied to groups of cells rather than groups of molecules. 'At such a stage,' Woese wrote, 'evolution was in effect communal.'[48]

Manfred Eigen expressed a similar idea in his 1970s papers on self-replicating hypercycles (see chapter 6). Eigen imagined a group of primitive organisms, each with a small set of genes. However, because the mechanisms for copying genes were error-prone, no one organism would have the necessary full set of genes. Eigen called this group of organisms a 'quasi-species'. Like Woese, he emphasised that they evolved as a group, not as individuals.

Some microorganisms may exist in this interdependent way even today. While there are millions of species of microbe, most have never been grown in a lab – and not for want of trying. Many have so far proved impossible to grow.[49] In some cases, the explanation seems to be that they cannot grow on their own. They must be paired with at least one other species from their habitat. Clearly, their need for each other is profound. In fact, some bacteria lack genes that were once thought to be essential, and may only be able to survive communally.[50]

This vision of the first life also resolves a long-standing question that we encountered in chapter 10: how did the first organisms sustain themselves? Wächtershäuser and Russell both argued that the first life was an autotroph that made its own building blocks: this was central to the vent hypothesis. In contrast, the pool hypothesis has the first organisms as heterotrophs whose building blocks were supplied by the surroundings.

In truth, the question is arguably meaningless. The first life was so intimately bound up with its surroundings that it is difficult to tell what should count as organism and what as surroundings. If chemical reactions in a pool make a protein, but this happens a millimetre away from the nearest RNA, which label should we use? We can only decide if we first choose which parts of the pool are alive and which aren't, and in the absence of a lipid membrane to define a cell, there is no good criterion. Instead, we could think of the entire pool as primitively alive. In that case it is an autotroph – unless key chemicals form outside and fall in, in which case it is arguably a heterotroph. This argument also holds for communities of protocells in a pool. Many key reactions will happen outside the protocells, or only in some of them with the products then being shared. Again, we cannot slap on a convenient label. One way or another, the building blocks of life were made, and whether we ascribe that synthesis to the first organisms or their environment is largely semantics.

Let's summarise. We have seen that the building blocks of life could all form from the same simple chemicals. This could have paved the way for chemoton-like protocells, like those Szostak has studied, to form on the early Earth. These protocells were surely clunky, but they would have survived as part of a co-operating group. This hypothesis takes all the best elements of earlier ideas like the RNA World, while sidestepping the chicken-and-egg paradoxes they throw up. It is our best answer to the question of how life began.

That leaves one final mystery: where did all this happen? In the last ten years a consensus has started to form that life began in hot, chemical-rich ponds on the first volcanic landmasses. Sutherland, Szostak, Deamer and others have all alighted on such an environment. Their ideas differ, but they all focus on small bodies of water on land.

This is disputed by people like Russell, who remain convinced that deep-sea alkaline vents are where life began. They argue there wasn't any land at first because the seas were too deep.

This is a perennial issue: origin-of-life hypotheses have often been contradicted by findings about the early Earth, as we saw in chapter 6. Many have found consolation in something Leslie Orgel supposedly said: 'Just wait a few years and conditions on the primitive Earth will change again.'[51] This is a little too cynical, but it does reflect the way our ideas have ping-ponged over the years.

However, there is growing evidence that dry land has existed since early in Earth's history. There were probably no large continents in the first two billion years of the planet's existence, because certain rocks only form in large continents and they are missing from strata over 2.5 billion years old.[52] However, no continents does not mean no land. The planet was volcanically active, so there were probably volcanic island chains like Hawaii.[53] Such landmasses seem to have been present 3.5 billion years ago. As we saw in chapter 1, the oldest confirmed microorganisms date from this time and are found in Australia. Recent analyses indicate they lived in a volcanic caldera dotted with hot springs – on land.[54]

Deamer believes the cradle of life was a geothermal pond: a pool of water that has been heated up by passing through hot rocks underground, and which is rich in chemicals.[55] Today, such ponds are found near Lassen Peak, an active volcano in California. There is a field of boiling springs and steam vents, reeking of sulphur. It is called Bumpass Hell, after a cowboy with the unfortunate name Kendall VanHook Bumpass, whose misfortunes escalated when he stepped in a pool of scalding-hot mud and lost a leg. Similar hot ponds can be found in Kamchatka in north-east Russia, around Mount Mutnowski.[56]

Such ponds have several apparent advantages. They will

experience cycles of wetting and drying, which was crucial for Deamer's experiments. They are exposed to sunlight, including ultraviolet radiation, which is useful for making biological chemicals: Sutherland's nucleotide synthesis relied on it. The idea is that cyanide-based chemicals would form in the air and react, giving rise to substances like amino acids and nucleotides. These would drift down into the ponds, where they could assemble into longer molecules like RNA and protein. The ponds had mineral surfaces – perhaps including montmorillonite – that could help the reactions along.

Deamer has repeatedly experimented with such pools. Visiting a geothermal pool in Kamchatka, he poured in a mix of simple biological chemicals like amino acids and the bases from nucleic acids.[57] Disappointingly, they did not self-assemble into longer chain molecules like RNA, probably because the water was acidic. Robert Shapiro, in one of his last articles, remarked that the study was 'a reminder that laboratory experiments don't always translate to nature'.[58]

However, the geothermal pools did better when Deamer tried to make protocells in them. For example, his team has added lipids to samples of water from the hot springs in Yellowstone National Park.[59] The lipids rapidly assembled into vesicles. In contrast, they did nothing of the sort in seawater, which is a strike against the idea that protocells formed in the sea. More recently, Deamer's team has made RNA-like chains within the protocells.[60]

More evidence in favour of geothermal ponds, and against the sea, was presented in 2012 by Armen Mulkidjanian: an Armenian-born biophysicist who has been thinking about the origin of life since the 1990s. He became involved after hearing about the alkaline vent hypothesis at a conference. 'I couldn't believe someone believed it,' he says. His sympathies lie with the soft version of the RNA World: the idea that replicating RNAs of some kind were at the core of the first life, alongside other components like lipid membranes.

Mulkidjanian's team asked a deceptively simple question.[61] What metals are found in cells, and how does that mixture compare to what is found in bodies of water on the Earth? The idea is that cells' internal makeup probably reflects the place where they formed. After all, the first cells cannot have had pumps to control what went in and out, so if they were floating in water with a particular metal dissolved in it, they would have picked up that metal.

The team found cells contain much more potassium than sodium. This was intriguing because some ancient proteins need potassium to function, but not sodium.[62] This preference for potassium over sodium is another strike against the first cells forming in seawater, which contains lots of sodium in the form of sodium chloride, or table salt. Instead, the evidence pointed to somewhere with more potassium than sodium, plus plenty of zinc, manganese and phosphate. Geothermal ponds, like those in Kamchatka, are the only places that fit.

Sutherland has proposed a subtly different alternative: streams flowing down the sides of a meteorite impact crater and meeting in a pool at the bottom.[63] He devised this after years of studying the cyanide chemistry that yielded nucleotides and other biological chemicals. A meteorite impact crater offered all the necessary chemicals. The impact would have formed hydrogen cyanide and thus cyanamide, the basis for all the syntheses. The meteorite itself would supply metals like iron and nickel, which were needed to speed up many of the reactions, and the crucial phosphate. And the whole scene would have been bathed in ultraviolet radiation, and could have undergone wet–dry cycles.

The reactions to make proteins, nucleic acids and lipids are different, and cannot take place in the same body of water. Sutherland's solution is to suppose that several rain-fed streams flowed down the crater sides.[64] This is hardly controversial, as anyone who has been hill-walking will know. 'You probably need to do it in four streams and then have those streams converge,' says Sutherland. Each stream would encounter different rocks and chemicals on its journey, and be in sun or shade at different times, pushing the reactions in different directions from the same starting point. 'They're variations on a theme rather than different pieces of music,' says Sutherland. Once all the building blocks had reached the pool at the bottom, they would self-assemble into something resembling Szostak's protocells, and life would have formed.

We do not yet have evidence to decide between these two locales. However, we do not have to choose yet. What is clear is that there is voluminous experimental evidence that the chemicals of life form in these terrestrial environments, and hints that they can self-assemble. Decades of experiments have shown that cyanide-based chemistry can provide the building blocks of life. Meanwhile, Szostak has shown that such building blocks can assemble into protocells that are startlingly lifelike. None of this has been shown for marine environments like deep-sea vents. Instead, the vent hypothesis demands a mechanism to convert carbon dioxide into the building blocks of life – and carbon dioxide is unreactive. Plants turn it into sugar using energy from sunlight, but this is complicated and the Last Universal Common

Ancestor probably could not do it. Worse, experimental simulations of alkaline vents have not produced anything terribly impressive.

For this reason, the notion that life began in a warm pool on land, using building blocks derived from cyanide reactions, seems by far the most promising. The ultimate test would be to build a living thing from scratch: to set up a Miller-type experiment simulating a primordial environment, which would run until something formed that was undeniably alive. It is a fool's game to predict when such an experiment might be performed. But it seems likely that it will be – and it is a good bet that the simulated environment will be a pond on land rather than the sea.

Even this dramatic demonstration would not wholly end the discussion. What if experiments ultimately show life can arise in more than one way? It would then be extremely difficult to decide which one really happened on Earth, or to rule out the possibility that several kinds of life arose in different places, and either merged or competed. The best we can hope to do in the near future is to show that one or more of the pathways creates new life.

If it turns out life really can form in a geothermal pond or a pool in a meteorite crater, one of the great ironies of scientific history will have occurred. For Charles Darwin would have got it right yet again, in a dashed-off letter that contained almost no detail. Rereading his 1871 missive now, it seems positively prophetic. Darwin envisioned life beginning in a 'warm little pond' (tick) that contained 'phosphoric salts' (an obsolete term for phosphate) and was bathed in 'light, heat, electricity'. Admittedly, he supposed that the first step towards life was the formation of 'a protein compound', which now seems simplistic.

But then Darwin wrote his letter in the same year that the world first learned of the existence of nucleic acids. The significance of these chemicals remained unclear, while the inner workings of cells were a mystery. Given the limited knowledge available to him, Darwin's guess was as prescient as it is possible to be.

Epilogue: The Meaning of Life

> 'Everything you know, your entire civilisation, it all begins right here in this little pond of goo. Appropriate somehow, isn't it?'
>
> *Star Trek: The Next Generation*, 'All Good Things . . .' written by Ronald D. Moore and Brannon Braga[1]

The previous fourteen chapters have told the story of the quest to understand how life began on Earth. The first hypothesis, the primordial soup as envisioned by Oparin and Haldane, ultimately failed because life turned out to be far more complex than either man knew. Later hypotheses like the RNA World tried instead to strip life down to one of its key components: either a chemical like a protein, or a process like metabolism. However, none of the experiments based on these ideas has produced something that is unambiguously alive, despite decades of effort. Instead, it turns out that the chemical building blocks of life all form from the same raw materials, and once formed have a tendency to self-assemble into structures that crudely resemble living cells. In other words, the first life was not based around a single component like a gene. Life began with all its major components, just in drastically simplified form.

For now, this is the closest we have come to explaining how life began on our planet. It has taken us the better part of a century to get this far, and the story is surely not over yet. There are many details to be fleshed out and problems to be solved. Nevertheless, I think the story as outlined above is likely to stick, for two reasons. First, it relies solely on chemicals that are either found in living organisms today, or are known to be common in nature: no artificial nucleic acids or semi-living clays are required. And second, it is rooted in some of the basic principles of biology and ecology; in particular, the fact that organisms always live in communities. The first organisms may well have been unfamiliar in their exact makeup, but in their fundamental nature they were much like us and every other living species. When we consider that each proto-organism was flawed and incomplete, able to do certain things but still needing its neighbours in order to survive, we can see ourselves in them.

Finally, let's consider three profound questions in the light of that finding. Is there extraterrestrial life? Would we know it if we saw it – or, in other words, what actually *is* life? And what does our knowledge of life's origin and nature mean for us?

First, what can we conclude about life elsewhere in the universe? Is it everywhere, or nowhere? Is the cosmos teeming with life, or a barren wasteland with Earth the only shining blue dot of vitality?

The first observation to make is that so far we have not found any hard evidence of extraterrestrial life. Consider the Solar System. Early in the twentieth century, many astronomers were confident that other planets had lush rainforests, maybe even intelligent civilisations. However, our forays into space have killed this idea. If there is life in the Solar System, it is probably single-celled and microscopic.

Mars is unquestionably the planet with the highest probability of life, which is not the same as saying that Martian life is likely. It has only a thin atmosphere and temperatures regularly fall far below 0°C. It is a wasteland of reddish rocks, scoured by dust storms. A few microorganisms have survived in simulated Martian conditions, but most die quickly.

When NASA's *Viking* landers reached Mars in 1976, they collected soil samples and tested them for signs of life. One experiment seemingly revealed life in the soil,[2] but the evidence was ambiguous and nowadays most agree it is not proof of life.[3] Similarly, a meteorite from Mars was hailed in the 1990s as containing preserved bacteria: 'fossil remains of a past Martian biota'.[4] But the evidence was again weak: a 2012 review favoured 'simple and down-to-earth' explanations for all of it.[5] For now the question of life on Mars remains unanswered. It might stay that way, as any microbes could be kilometres underground.

However, even if Mars is not habitable now, it may have been when it was young. The surface is streaked with valleys, probably carved by running water.[6] There is liquid water under the southern polar ice cap.[7] Mars was also volcanically active in the past, so may have had geothermal pools where life could have started.

The other promising targets are moons orbiting giant planets: Europa, one of the moons of Jupiter, and Saturn's moons Enceladus and Titan.

Europa and Enceladus are almost twins. Both look like dirty white snooker balls. Their surfaces seem to be mostly ice, but astronomers strongly suspect they have liquid oceans under the ice.[8] Both have been seen venting plumes of water vapour hundreds of kilometres high.[9] The presence of oceans marks them as arguably more similar to Earth than any known world.[10] However, hypotheses about life on Europa and Enceladus almost always revolve around a hydrothermal vent on the seabed – and, as we have seen, it is far from clear that life can originate in such vents.

Finally, there is Titan: the only moon with a thick atmosphere. It is intensely cold: the temperature has been estimated at minus 179°C. Despite these extreme conditions, Titan has lakes and oceans. The biggest, Kraken Mare,* spans 400,000 square kilometres: larger than the Caspian Sea.[11] The twist is that Titan's oceans are not made of water.[12] Instead, they are made of simple carbon-based chemicals like methane. These are normally gases on Earth but have condensed into liquid in the bitter cold. There may also be another ocean hidden underground.[13]

Titan is essentially a giant prebiotic laboratory.[14] It holds many of the chemicals that have cropped up during efforts to make biological molecules, including cyanides.[15] For this reason, many astrochemists suspect complex molecules like RNA could exist there.[16] Any life would be profoundly alien, because it would not have formed in liquid water.[17] However, the cold will surely slow down the necessary chemical reactions. It remains to be seen whether that is a roadblock or a deal-breaker.

The study of the origin of life suggests that Mars, which once

* Maps of Titan are particularly entertaining for a certain type of person because, instead of naming features after ancient myths and legends, astronomers have drawn many of the names from J. R. R. Tolkien's fictional Middle-Earth. As a result, one of Titan's tallest mountains is called Doom Mons.

had volcanoes and flowing water, is the most likely place to find alien life. In particular, if life truly begins in surface pools, Europa and Enceladus cannot be cradles of life.[18]

What about life beyond the Solar System? So far there is no sign of it. Since the 1960s, researchers have scanned the skies for signals from aliens, to no effect. A project called Breakthrough Listen eavesdropped on 1327 stars for four years, and found nothing resembling a signal.[19] There is also no evidence of alien engineering projects, such as a Dyson Swarm: a cloud of artificial structures that advanced aliens might build around a star to harness all its energy.* In 2015 astronomers announced that such a thing might have been found orbiting a star called KIC 8462852 – now known as 'Tabby's Star' after lead researcher Tabetha Boyajian. The star had dimmed up to 20 per cent for days at a time.[20] One possible explanation was a Dyson Swarm, but it was probably comets and dust.[21]

What are we to make of the apparent absence of intelligent extraterrestrials? It is odd, as presumably the Earth is not especially unusual. Are the aliens communicating using a technology that we cannot access, hibernating until the universe has cooled down, or hiding from malevolent swarms of killer robots?† Alternatively, are human-like levels of intelligence not a successful strategy? In just a few millennia, humans have overheated our planet, threatened much of the wildlife on which we depend, and chosen a succession of bloodthirsty narcissists to have control

* This is a variant on a Dyson Sphere, which was mentioned in chapter 7. Both would entirely surround a star, but the Sphere is imagined to be a single vast structure, while the Swarm would be made up of many smaller objects.
† These are among the more sensible suggestions.

over arsenals of nuclear weapons that could wipe us out in a day. Maybe 'intelligent' civilisations kill themselves off quickly.

However, our failure to find intelligent life says nothing about the prevalence of life in general. We have only had radio for a little over a century, whereas life has existed on Earth for 3.5 billion years. If Earth's history is representative, most life is single-celled, so microbes are what we should expect to find.

Where could they live? Since 1992 astronomers have found over 4000 planets around other stars.[22] However, most look uninhabitable. They are 'hot Jupiters': gas giants that orbit close to their stars and are swelteringly hot. Instead, planets must be in the 'habitable zone': close enough to a star that liquid water can exist, but not so close that temperatures get too hot for life. The planet probably also has to be the right sort of size. That means similar to Earth, but we do not know how similar. Many planets also have wild orbits, swinging further away from their stars and then closer in, so their surface temperatures must change dramatically.[23] There is also the question of whether a habitable planet needs a large-ish moon to stabilise its motion, as Earth does.*

Again, studies of the origin of life offer critical clues. In 2018 a group that included John Sutherland tried to define the region around a given star where life could form – as opposed to simply survive, which is what the habitable zone tells us.[24] They argued that ultraviolet radiation is necessary to make the molecules of life, so life-generating planets had to be close enough to their

* It's a bit of a spoiler to say this, but N. K. Jemisin's brilliant *The Fifth Season* and its sequels employ this idea.

star to get a good dose. It turned out that many exoplanets are in their star's habitable zone but outside the life-generating zone, suggesting they are sterile. Only eight known exoplanets lie in both zones, seven of which look too big.[25] The one possible is Kepler-452b, which is 1.6 times the size of Earth. Inconveniently, it is 1400 light-years away.

The fact is that we do not know if life is common on other planets, vanishingly rare, or something in between. Our intuitions will not help us. But studying the origin of life will: if we find a process by which non-living matter assembles itself into an organism, we can study that process. Which temperature does it need? Which substances have to be present, or absent? The answers will narrow the search for living worlds.

However, this raises our second question. If we ran across a living organism on another world, would we recognise it? Put another way, what is this life thing anyway?

Specifying what life is seems easy. An enraged male elephant is definitely alive and a granite boulder is not. But as we've seen, there are many borderline cases. Most biologists agree that viruses are not alive, but they are also not non-living in the way that a brick is non-living. The protocells and self-replicating chemical mixtures created by origin-of-life researchers blur the line further. Similarly, cars have many of the traits we associate with life – they move and must be 'fed' – but they are not alive.

For this reason, there is no definition of life that everyone agrees on. Not for want of trying: a 2012 study found 123 distinct definitions.[26] For instance, many of us were taught that you could sum up life with the mnemonic 'MRS GREN', meaning 'movement, respiration, sensitivity, growth, reproduction, excretion and nutrition'. All living things are supposed to have these seven properties. However, it is easy to find exceptions. For example, worker honeybees and post-menopausal women do not reproduce, but they are firmly alive.

Another approach focuses on life's intimate relationship with the Earth. Volcanoes and alkaline vents are mechanisms for shuttling energy and matter between Earth's interior and the surface. If one of these transfer zones is the cradle of life, then life is Earth's way of balancing itself out, ensuring that things like electrons are evenly shared between layers.[27] There is surely some truth in this, as life is closely bound up with the Earth. But this is less a definition of life and more a description of its environment.

A more promising approach was set out by biochemist Addy Pross in his 2012 book *What Is Life?*,* in which he says living organisms all have 'dynamic kinetic stability'.[28] This is a mouthful, but it means something simple: all living organisms are simultaneously changing and static. Within your body, your heart is pumping blood and the interiors of your cells are frantically active. And yet, this activity mostly keeps your body the same. Unless something dramatic has happened, your body is the same shape today as it was yesterday. Furthermore, if you have children they will look like you. It is this interplay between change and stasis that for Pross lies at the heart of life.

Pross also thinks replicating molecules like DNA are essential for life, because they drive other molecules towards dynamic

* Confusingly, this is the same title as Erwin Schrödinger's classic book, although it does have a different subtitle.

kinetic stability. He therefore calls life 'a self-sustaining kinetically stable dynamic reaction network derived from the replication reaction'. It is hard to find exceptions to this. Some non-living phenomena have dynamic kinetic stability, like whirlpools in the sea, but they don't have self-replicating molecules.

Finally, there is the most famous definition. It was drawn up in 1994 by the Exobiology Discipline Working Group: a NASA committee that included biochemist Gerald Joyce, which advised NASA on extraterrestrial biology. They decided to create a definition of life, just so they could specify what they were talking about.[29] Joyce later included it in his foreword to a book, and it has been widely discussed.[30]

The NASA definition is: 'Life is a self-sustaining chemical system capable of Darwinian evolution.' It largely matches Pross's definition, at least conceptually. Both emphasise life's ability to stay the same. Similarly, NASA's point about life being 'capable of Darwinian evolution' corresponds to Pross's emphasis on 'a replication reaction': it is just that the NASA definition emphasises evolution, while Pross highlights the molecules that underlie evolution. However, it is unclear whether 'system' refers to an individual organism or a group.[31] This is a problem: an individual rabbit is alive, but cannot undergo Darwinian evolution without another rabbit. The Pross definition sidesteps this.

One problem with all these definitions is that they are only based on Earth life and we do not know how much life can vary. Does life need DNA? Does it need something *like* DNA? Must it be based on carbon and water, or are other chemistries possible? Our definitions must bear this in mind.[32]

However, there is also a deeper problem. There may not be a

hard divide between the living and the non-living. Instead, we should consider the possibility that the universe doesn't work like that. Maybe life is a concept we are imposing on reality, not something that can be objectively defined. In contrast, we can give a strict definition of an electron that works every time. That suggests electrons are a real, distinct phenomenon. But life may be less distinct than electrons. Consider the spectrum of human behaviours. We often make decisions like 'this is neurotypical but that is autistic' or 'this person enjoys drinking alcohol but that person is an alcoholic'. These differences exist, but where we draw the line is a value judgement. Deciding whether something is alive may be equally subjective.

This is not a new idea: many familiar concepts are fuzzy. In *A Modern Utopia*, the science fiction writer H. G. Wells asked readers to consider the word 'chair':*

> When one says chair, one thinks vaguely of an average chair. But collect individual instances, think of armchairs and reading chairs, and dining-room chairs and kitchen chairs, chairs that pass into benches, chairs that cross the boundary and become settees, dentists' chairs, thrones, opera stalls, seats of all sorts, those miraculous fungoid growths that cumber the floor of the Arts and Crafts Exhibition, and you will perceive what a lax bundle in fact is this simple straightforward term.

* Ludwig Wittgenstein explored similar ideas in *Philosophical Investigations*. However, one of these writers is easier to read than the other.

Maybe we should embrace this fuzziness. Some researchers now argue that it is 'impossible to define a "natural" frontier between non-living and living systems'.[33] Rather, we could decide if things are alive using fuzzy logic, which abandons straightforward true/false dichotomies and instead allows statements to have varying degrees of truth. Instead of 'alive' or 'not alive', we could devise a scale of aliveness.[34] Sutherland, Pross and a colleague have applied this idea to the origin of life.[35] They say that 'life would have emerged stepwise, through states of partial "aliveness", rather than through some single sharp transition'. The intermediate states might be called 'almost life'.[36] Today viruses would fall into this grey area, as would constructs like Szostak's protocells.[37]

This gradualistic approach seems far more plausible than any attempt to draw a hard dividing line. A fully alive organism would meet all the criteria of the NASA or Pross definitions of life, while a partially alive proto-organism would not.

Finally, what does our new understanding of life mean for us?

The first lesson is that life probably won't last, even setting aside our tendency towards self-destruction. Cosmologists suspect that there will not be many more generations of stars, because the matter in the universe is spreading out. Before too many more billions of years have passed, the matter will be too thinly spread to ever condense into a new star. The universe will become colder and darker, and any rocky planets will be less active and thus unlikely to give rise to life. On this view, we exist in a special period in the universe's history: the one short epoch where isolated pockets of life can emerge, before everything becomes fundamentally uninteresting.

This is a grim and deeply pessimistic view of our future. Cosmology seems to tell us that, in the very long run, we are doomed. Of course, we can speculate about fantastical future technologies, like diving into wormholes to escape to another, younger universe. But maybe it would be better to accept our ultimate mortality and consider how to live with it. There's a curiously profound line of dialogue in *Avengers: Age of Ultron*, a film that's otherwise a bit of a mess. The android character Vision is told that the human race is destined to become extinct. He replies: 'Yes. But a thing isn't beautiful because it lasts.' If humanity is only going to be here for a little while, we could at least become something kind and learned enough to be proud of before we depart.

When people are told about the theory of evolution and that they are descended from apes, fish and ultimately bacteria, they often react badly. It seems so humbling, so dirty and animalistic, that we should essentially be a modified chimpanzee. The fact we seem to have had our origins in some kind of primordial ooze near a volcano, reeking of hydrogen sulphide, only compounds this.

But there is another way of looking at it. The chemicals of life are plentiful, because they were born in the hearts of stars, and the stars are everywhere. The assembly of those substances into living cells is a localised reversal of the Second Law of Thermodynamics, a process that was enabled and shaped by the little blue-green planet on which it took place. On every level, our origins are deeply connected with the fundamental building blocks and laws of the universe. To again quote Carl Sagan: 'We are a way for the cosmos to know itself.' We might be small but we are a part of the indivisible whole.

Arguably, we are the most important thing in the universe (unless other sentient beings exist, in which case we should treat them as equals). If we were not here to live in the universe and marvel at its beauty, what would be the point of any of it? It is only living beings that are conscious, and therefore we are the only means by which the universe can ever be ascribed value or meaning. There is no purpose in inanimate matter and energy, unless sentient beings like us choose to see it. I am fond of a passage by the German writer Johann Wolfgang von Goethe,* describing the sublime pleasure of feeling at one with the universe:[38]

> When the healthy nature of man functions as a totality, when he feels himself in the world as in a vast, beautiful, worthy and valued whole, when a harmonious sense of well-being affords him pure and free delight – then the universe, if it were capable of sensation, would exult at having reached its goal, and marvel at the culmination of its own development and being. For what is the use of all the expenditure of suns and planets and moons, of stars and galaxies, of comets and nebulae, of completed and developing worlds, if at the end a happy man does not unconsciously rejoice in existence?

* This passage is a little obscure. It comes from an 1805 essay about the art historian and archaeologist Johann Winckelmann. I first ran across it in the introduction to my copy of Nietzsche's *Thus Spoke Zarathustra*, and loved it immediately.

Our links to the wider cosmos may seem abstract and remote, but our connections to the Earth and the rest of its biosphere are much closer and more intimate. All living organisms are quite literally our distant cousins. Even the most unfamiliar bacteria are our relatives, for we all descend from the same ancestral population. We are also absolutely reliant on all of them for our survival and happiness. Plants and photosynthetic microorganisms pump out the oxygen we breathe as a waste product. They are also the source of all the food we eat: some of us eat animals, but those animals only lived because of plants. Pollinating insects ensure that the crops we grow are fertilised. Other organisms do not have such obvious benefits, but ecologists have found that ecosystems with a greater range of species are more stable, and therefore safer for us to live in. Finally, simply visiting a park, forest or wilderness helps to restore our mental wellbeing.

Addy Pross has asked a truly challenging question: 'Do individual life forms actually exist?'[39] The answer seems obvious, but when you consider how interdependent organisms are, both on members of their own species and on the wider ecosystem, our apparent individuality starts to look distinctly hazy. Our intimate reliance on the ecosystem becomes clear if you try to envision a spaceship suitable for interstellar travel. Consider the fantastic technologies that would be needed to ensure a supply of oxygen, clean water, food and a stable environment. Then ask yourself, would it ultimately be easier to create a microcosm of Earth; a miniature biosphere in which to make the journey?*

* This idea is explored in Kim Stanley Robinson's deeply moving novel *Aurora*.

Our connections to the planet itself are similarly profound. Its atmosphere ensures that the surface never gets too hot or cold for life (notwithstanding our continuing greenhouse gas emissions, which are heating up the climate to a dangerous extent). Earth's magnetic field, powered by the core of the planet, protects us from the worst of the Sun's radiation. Our deepest origins are completely interwoven with the planet and its processes: with rocks, minerals and water.

It is generally a bad idea to draw out a moral from science, because it only tells us how the world is, which is not necessarily how it should be. But I do think there is a fundamental lesson in the way life began, and it is this. The Earth is our cradle, our home and our parent. We have not yet found anywhere else that remotely compares. We live here, or nowhere. If we want to thrive on Earth, to be in turn happy and energetic and creative and tranquil, then it seems to me we have no choice but to be kind. We must care for each other, tend to the distant relatives with whom we coexist, and preserve the little blue planet that nurtures us all. Even when everything else about our society threatens to tear us apart, our living planet can bind us together.

ACKNOWLEDGEMENTS

This book would not exist if it were not for the support of hundreds of people: both during the writing, and in the years leading up to it.

Thanks to the many people who agreed to share their recollections of the key events. They are: Jeffrey Bada, Donna Blackmond, Dorothy 'Dodo' Cairns-Smith, H. James Cleaves, David Deamer, Ernesto Di Mauro, James Kasting, Antonio Lazcano, Armen Mulkidjanian, Eric Parker, Michael Russell, Alan Schwartz, John Sutherland and Günter Wächtershäuser.

Many of these people were also kind enough to critically review various chapters, for which a second thank-you is owed. I owe a particular debt to Matthew Cobb, who kindly reviewed chapter 4 after I emailed him out of the blue. I also want to thank two anonymous reviewers, who offered useful critiques of the book as a whole. Any remaining errors are solely mine.

I also owe a great debt to my panel of friendly non-specialist readers, who read the book and told me which bits were dull or incomprehensible. Without Ian Maun, Lindsey Brown,

Joost van Es and Sarah Marshall-Maun, it would be a lot less entertaining. Any remaining dullness or incomprehensibility is my fault.

I'm profoundly grateful to my agent Peter Tallack of Science Factory, who shepherded the book from its inception through to publication and was an endless fount of calm wisdom. Paul Murphy signed me to Weidenfeld & Nicolson but sadly left before I completed the book. However, Maddy Price took it on with enormous enthusiasm when she joined the company a few months later, and I've been grateful for her constant encouragement. She proposed a number of key changes that enormously improved the final result. Rosie Pearce and her crack team of copy editors and lawyers also made a big difference. In the US, Karen Merikangas Darling of the Chicago University Press has been equally supportive and also arranged for the anonymous reviews.

I've been supported over a decade of science journalism, much of it focused on the origin of life, by a small army of colleagues. There are too many to name individually, but I am especially grateful to Colin Barras, Adam Becker, Catherine Brahic, Andy Coghlan, Niall Firth, Melissa Hogenboom, Rowan Hooper, Michael Le Page, Pierangelo Pirak, Sumit Paul-Choudhury, Penny Sarchet, Kara Šegedin, Matt Walker and Emily Wilson.

The final thank-yous are to my awesome wife Sarah and our equally excellent daughter Libbet, a source of endless joy. Also, Sarah came up with the joke about the Schrödinger's cat thought experiment. No Sarah and Libbet, no book.

BIBLIOGRAPHY

Cairns-Smith, A. G. *Seven Clues to the Origin of Life*. 1985. Cambridge University Press.

Clark, R. *J. B. S.: The Life and Work of J. B. S. Haldane*. 1968. Hodder and Stoughton Limited.

Cobb, M. *Life's Greatest Secret*. 2015. Profile Books.

Cockell, C. *The Equations of Life: How physics shapes evolution*. 2018. Basic Books.

Davies, P. *The 5th Miracle: The search for the origin and meaning of life*. 1999. Simon & Schuster.

Deamer, D. W. *Assembling Life: How can life begin on Earth and other habitable planets?* 2019. Oxford University Press.

De Duve, C. *Vital Dust: Life as a cosmic imperative*. 1994. BasicBooks, a division of HarperCollins Publishers.

Fox, S. W. *The Emergence of Life: Darwinian evolution from the inside*. 1988. Basic Books.

Fry, I. *The Emergence of Life on Earth: A historical and scientific overview*. 2000. Rutgers University Press.

Gánti, T. *Az élet princípiuma* (*The Principles of Life*). 1971. Gondolat, Budapest.

Gánti, T. *A Theory of Biochemical Supersystems*. 1979. Akadémiai Kiadó, Budapest.

Grand, S. *Creation: Life and how to make it*. 2000. Weidenfeld & Nicolson.

Ings, S. *Stalin and the Scientists: A history of triumph and tragedy 1905–1953*. 2016. Faber and Faber Ltd.

Kauffman, S. A. *The Origins of Order*. 1993. Oxford University Press.

Lane, N. *The Vital Question: Why is life the way it is?* 2015. Profile Books Ltd.

Mesler, B.; Cleaves, H. J. *A Brief History of Creation: Science and the search for the origin of life*. 2016. W. W. Norton & Company, Inc.

Monod, J. *Le hazard et la nécessité*. 1970. Éditions de Seuil, Paris. Published in English as *Chance and Necessity* by William Collins Sons & Co Ltd, 1972.

Morowitz, H. J. *Energy Flow in Biology: Biological organization as a problem in thermal physics*. 1968. Academic Press, Inc.

Morowitz, H. J. *Beginnings of Cellular Life: Metabolism recapitulates biogenesis*. 1992. Yale University Press.

Mukherjee, S. *The Gene: An intimate history*. 2016. Vintage, Penguin Random House.

Prebble, J.; Weber, B. *Wandering in the Gardens of the Mind: Peter Mitchell and the making of Glynn*. 2003. Oxford University Press.

Pross, A. *What Is Life? How chemistry becomes biology*. 2012. Oxford University Press.

Reynolds, J. A.; Tanford, C. *Nature's Robots: A History of Proteins* (Oxford Paperbacks). 2003. Oxford University Press.

Rutherford, A. *Creation: The origin of life/The future of life*. 2013. Viking.

Scharf, C. *The Copernicus Quest: The quest for our cosmic (in)significance*. 2014. Allen Lane.

Schrödinger, E. *What Is Life? The physical aspect of the living cell*. 1944. Cambridge University Press.

Shapiro, R. *Origins: A skeptic's guide to the creation of life on Earth*. 1986. Summit Books.

Woese, C. R. *The Genetic Code: The molecular basis for genetic expression*. 1967. Harper & Row.

NOTES

Introduction

1 Monod, J. *Le hazard et la nécessité*. 1970. Éditions de Seuil, Paris. Published in English as *Chance and Necessity* by William Collins Sons & Co Ltd, 1972.
2 Blobel, G. 'Christian de Duve (1917–2013).' *Nature*, volume 498, issue 7454, page 300. 2013. https://doi.org/10.1038/498300a
3 De Duve, C. 'Life as a cosmic imperative?' *Philosophical Transactions of the Royal Society A: Mathematical, Physical and Engineering Sciences*, volume 369, issue 1936, pages 620–623. 2011. https://doi.org/10.1098/rsta.2010.0312
4 Saini, A. Inferior: *How science got women wrong and the new research that's rewriting the story*. 2017. HarperCollins.

PART ONE: PRIMORDIAL SCIENCE

1 Published as: Whitesides, G. M. 'Revolutions in chemistry.' *Chemical & Engineering News*, volume 85, number 13, pages 12–17. 2007. https://pubs.acs.org/cen/coverstory/85/8513cover1.html

Chapter 1: The Biggest Question

1 Gaiman, N. *Norse Mythology*. 2017. Bloomsbury.
2 Wöhler, F. 'Uber künstliche Bildung des Harnstoffs.' *Annalen der Chemie und Pharmacie*, volume 12, pages 253–256. 1828. https://doi.org/10.1002/andp.18280880206

3 Ramberg, P. J. 'The Death of Vitalism and The Birth of Organic Chemistry: Wohler's Urea Synthesis and the Disciplinary Identity of Organic Chemistry.' *Ambix*, volume 47, issue 3, pages 170–195. 2000. http://dx.doi.org/10.1179/amb.2000.47.3.170
4 Moore, B. *The Origin and Nature of Life*. 1913. Henry Holt and Company (New York), Williams and Norgate (London).
5 Grand, S. *Creation: Life and how to make it*. 2000. Weidenfeld & Nicolson. Page 6.
6 Fry, I. *The Emergence of Life on Earth: A historical and scientific overview*. 2000. Rutgers University Press.
7 Pouchet, F. 'Note sur des proto-organismes végétaux et animaux, nés spontanément dans l'air artificiel et dans le gaz oxygène', *Comptes Rendus*, volume 47, pages 979–984, Académie des Sciences, 1858.
8 Pouchet, F. *Hétérogénie, ou Traité de la génération spontanée*. 1859. Paris: Baillière.
9 Darwin, C. *On the Origin of Species by Means of Natural Selection, or the Preservation of Favoured Races in the Struggle for Life*. 1859. John Murray, London.
10 Mesler, B.; Cleaves, H. J. *A Brief History of Creation: Science and the search for the origin of life*. 2016. W. W. Norton & Company, Inc.
11 https://www.darwinproject.ac.uk/letter/DCP-LETT-7471.xml
12 Hooke, R. *Micrographia: or Some Physiological Descriptions of Minute Bodies Made by Magnifying Glasses. With Observations and Inquiries Thereupon*. 1665. Royal Society, London.
13 Van Leeuwenhoek, Antonie. 'Observations, communicated to the publisher by Mr. Antony van Leewenhoeck, in a dutch letter of the 9th Octob. 1676. here English'd: concerning little animals by him observed in rain-well-sea- and snow water; as also in water wherein pepper had lain infused.' *Philosophical Transactions of the Royal Society*, volume 12, number 133, pages 821–831. 25 March 1677. http://dx.doi.org/10.1098/rstl.1677.0003
14 Schleiden, Matthias Jacob. 'Beiträge über Phytogenesis.' *Müller's Archiv für Anatomie and Physiologie*, pages 137–176. 1838. English translation online at https://www.biodiversitylibrary.org/item/43441#page/291/mode/1up
15 Schwann, T. *Microscopical Researches into the Accordance in the Structure and Growth of Animals and Plants*. 1839. Translated into English 1847. http://vlp.mpiwg-berlin.mpg.de/library/data/lit28715/index_html?pn=7&ws=1.5
16 Baker, J. R. 'The Cell-theory: a Restatement, History, and Critique. Part IV. The Multiplication of Cells.' *Quarterly Journal of Microscopical Science*, volume 94, issue 4, pages 407–440. 1953. http://jcs.biologists.org/content/s3-94/28/407

17 Lagunoff, D. 'A Polish, Jewish Scientist in 19th-Century Prussia.' *Science*, 20 December 2002, volume 298, issue 5602, page 2331. http://dx.doi.org/10.1126/science.1080726
18 Ussher, J. *Annales Veteris Testamenti, a prima mundi origine deducti, una cum rerum Asiaticarum et Aegyptiacarum chronico, a temporis historici principio usque ad Maccabaicorum initia producto*. 1650.
19 Gould, S. J. 'Fall in the House of Ussher.' *Natural History*, volume 100, pages 12–21. November 1991. http://www.stephenjaygould.org/library/gould_house-ussher.html Also republished in *Eight Little Piggies*, W. W. Norton & Company, 1993.
20 Thomson, W. 'On the Secular Cooling of the Earth.' *Transactions of the Royal Society of Edinburgh*. XXIII, pages 160–161. 1864.
21 Stacey, F. D. 'Kelvin's age of the Earth paradox revisited.' *Journal of Geophysical Research: Solid Earth*, volume 105, issue B6, pages 13155–13158. https://doi.org/10.1029/2000JB900028
22 Boltwood, B. B. 'Ultimate Disintegration Products of the Radio-active Elements. Part II. The disintegration products of uranium.' *American Journal of Science*, series 4, volume 23, pages 77–88. 1907. https://doi.org/10.2475/ajs.s4-23.134.78
23 Dunham, K. C. 'Arthur Holmes 1890–1965.' *Biographical Memoirs of Fellows of the Royal Society*, volume 12, pages 290–310. 1966. https://doi.org/10.1098/rsbm.1966.0013
24 Holmes, A. 'The Association of Lead with Uranium in Rock-Minerals, and Its Application to the Measurement of Geological Time.' *Proceedings of the Royal Society A: Mathematical, Physical and Engineering Sciences*, volume 85, issue 578, page 248. 1911. https://doi.org/10.1098/rspa.1911.0036
25 Holmes, A. *The Age of the Earth*. 1913. Harper & Brothers.
26 Holmes, A. 'An estimate of the age of the Earth.' *Nature*, volume 157, pages 680–684. 1946. https://doi.org/10.1038/157680a0
27 Houtermans, F. G. 'Determination of the age of the Earth from the isotopic composition of meteoritic lead.' *Nuovo Cimento*, volume 10, issue 12, pages 1623–1633. 1953. https://doi.org/10.1007/BF02781658
28 Patterson, C. C. 'The isotopic composition of meteoritic, basaltic and oceanic leads, and the age of the Earth.' *Proceedings of the Conference on Nuclear Processes in Geologic Settings*, Williams Bay, Wisconsin, Sept. 21–23, 1953, pages 36–40.
29 Patterson, C. C. 'Age of meteorites and the Earth.' *Geochimica et Cosmochimica Acta*, volume 10, pages 230–237. 1956.
30 Dalrymple, G. B. 'The age of the Earth in the twentieth century: a problem (mostly) solved.' *Special Publications, Geological Society of London*, volume 190, issue 1, pages 205–221. 2001. http://www.blc.arizona.edu/

courses/schaffer/449/Geology/Dalrymple%20Geol%20Time.pdf

31 Hartmann, W. K.; Davis, D. 'Satellite-sized planetesimals and lunar origin.' *Icarus*, volume 24, pages 504-515. 1975. https://doi.org/10.1016/0019-1035(75)90070-6

32 Ford, T. D. 'Precambrian fossils from Charnwood Forest'. *Yorkshire Geological Society Proceedings*, volume 31, issue 3, pages 211–217. 1958. https://doi.org/10.1144/pygs.31.3.211

33 Sprigg, R. C. 'Jellyfish from the Basal Cambrian in South Australia.' *Nature*, volume 161, pages 568–569. 1948. https://doi.org/10.1038/161568a0

34 Lowe, D. R. 'Stromatolites 3,400-Myr old from the Archean of Western Australia.' *Nature*, volume 284, issue 5755, pages 441–443. 1980. https://doi.org/10.1038/284441a0
—Walter, M. R.; Buick, R.; Dunlop, J. S. R. 'Stromatolites 3,400–3,500 Myr old from the North Pole area, Western Australia.' *Nature*, volume 284, issue 5755, pages 443–445. 1980. https://doi.org/10.1038/284443a0

35 Abramov, O.; Mojzsis, S. J. 'Microbial habitability of the Hadean Earth during the late heavy bombardment.' *Nature*, volume 459, pages 419–422. 2009. https://doi.org/10.1038/nature08015

36 Boehnke, P.; Harrison, T. M. 'Illusory Late Heavy Bombardments.' *Proceedings of the National Academy of Sciences*, volume 113, issue 39, pages 10802–10806. 2016. https://doi.org/10.1073/pnas.1611535113

37 Lowe, D. L.; Byerly, G. R.; Kyte, F. T. 'Recently discovered 3.42–3.23 Ga impact layers, Barberton Belt, South Africa: 3.8 Ga detrital zircons, Archean impact history, and tectonic implications.' *Geology*, volume 42, issue 9, pages 747–750. 2014. https://doi.org/10.1130/G35743.1

38 Dodd, M. S.; Papineau, D.; Grenne, T.; Slack, J. J.; Rittner, M.; Piraino, F.; O'Neil, J; Little, C. T. S. 'Evidence for early life in Earth's oldest hydrothermal vent precipitates.' *Nature*, volume 543, pages 60–64. 2 March 2017. https://doi.org/10.1038/nature21377

39 Bell, E. A.; Boehnke, P.; Harrison, T. M.; Mao, W. L. 'Potentially biogenic carbon preserved in a 4.1 billion-year-old zircon.' *Proceedings of the National Academy of Sciences*, volume 112, issue 47, pages 14518–14521. November 2015. http://dx.doi.org/10.1073/pnas.1517557112

Chapter 2: A Soviet Free Thinker

1 Ings, S. *Stalin and the Scientists: A history of triumph and tragedy 1905–1953*. 2016. Faber and Faber Ltd.

2 Oparin, A. I. *Proiskhozhdenie zhizni*. Moscow: Izd. Moskovskii Rabochii, 1924. English translation: https://www.valencia.edu/orilife/textos/The%20Origin%20of%20Life.pdf

3 Miller, S.; Schopf, J.; Lazcano, A. 'Oparin's "Origin of Life": Sixty years

later.' *Journal of Molecular Evolution*, volume 44, issue 4, pages 351–353. 1997. https://doi.org/10.1007/PL00006153

4 Clark, R. *J. B. S.: The Life and Work of J. B. S. Haldane*. 1968. Hodder and Stoughton Limited.

5 Haldane, J. B. S. 'The origin of life.' *Rationalist Annual*, volume 148, pages 3–10. 1929. https://www.uv.es/~orilife/textos/Haldane.pdf

6 Oparin, A. I. *Vozniknovenie zhizni na zemle*. Moscow: Izd. Akad. Nauk SSSR, 1936. http://krishikosh.egranth.ac.in/bitstream/1/2027792/1/164418.pdf

7 Piacentini, E. 'Coacervation.' In: Drioli E.; Giorno L. (eds) *Encyclopedia of Membranes*. 2016. Springer, Berlin, Heidelberg. https://doi.org/10.1007/978-3-662-44324-8

8 Mukherjee, S. *The Gene: An intimate history*. 2016. Vintage, Penguin Random House.

9 Gardner, M. *Fads and Fallacies in the Name of Science*. 1957. Dover Publications Inc.

10 Graham, L. R. *Science in Russia and the Soviet Union: A short history*. 1993. Cambridge University Press.

11 Fox, S. W. (ed) 'The Origins of Prebiological Systems and of their Molecular Matrices: Proceedings of a conference conducted at Wakulla Springs, Florida, on 27–30 October 1963 under the Auspices of the Institute for Space Biosciences, the Florida State University and the National Aeronautics and Space Administration.' 1965. Elsevier, Inc.

12 Haldane, J. B. S. 'Cancer's a Funny Thing.' *New Statesman*, 21 February 1964, page 298.

Chapter 3: Creation in a Test Tube

1 'Harold Clayton Urey: Biographical.' From *Nobel Lectures, Chemistry 1922–1941*, Elsevier Publishing Company, Amsterdam. 1966. https://www.nobelprize.org/prizes/chemistry/1934/urey/biographical/

2 Arnold, J. A.; Bigeleisen, J.; Hutchison, C. A. 'Harold Clayton Urey 1893–1981.' In: *Biographical Memoirs*, volume 68. 1995. Washington, DC: The National Academies Press. https://www.nap.edu/read/4990/chapter/19

3 Cohen, K. P.; Runcorn, S. K.; Suess, H. E.; Thode, H. G. 'Harold Clayton Urey, 29 April 1893 – 5 January 1981.' *Biographical Memoirs of Fellows of the Royal Society*, volume 29, pages 622–659. 1983. https://doi.org/10.1098/rsbm.1983.0022

4 Urey, H. C.; Brickwedde, F. G.; Murphy, G. M. 'A Hydrogen Isotope of Mass 2.' *Physical Review*, volume 39, issue 1, pages 164–165. 1932. https://dx.doi.org/10.1103/PhysRev.39.164

5 Garrison, W. M.; Morrison, D. C.; Hamilton, J. G.; Benson, A. A.;

Calvin, M. 'Reduction of carbon dioxide in aqueous solutions by ionizing radiation.' *Science*, volume 114, issue 2964, pages 416–418. 1951. https://doi.org/10.1126/science.114.2964.416

6 Bada, J. L.; Lazcano, A. 'Stanley L. Miller, 1930–2007: a biographical memoir.' In: *Biographical Memoirs*. 2012. Washington, DC: The National Academies Press. http://www.nasonline.org/member-directory/deceased-members/52506.html

7 Bada, J. L.; Lazcano, A. 'Stanley Miller's 70th birthday.' *Origins of Life and Evolution of the Biosphere*, volume 30, issue 2–4, pages 107–112. 2000. https://doi.org/10.1023/A:1006746205180

8 Interview with Antonio Lazcano.

9 Miller, S. L. 'A production of amino acids under possible primitive Earth conditions.' *Science*, volume 117, issue 3046, pages 528–529. 1953. http://dx.doi.org/10.1126/science.117.3046.528

10 'Science: Semi-creation.' *Time*, volume LXI, number 21, 25 May 1953. http://content.time.com/time/magazine/article/0,9171,890596,00.html

11 'Life and a glass Earth.' *New York Times*, 17 May 1953, page 10. https://www.nytimes.com/1953/05/17/archives/life-and-a-glass-earth.html

12 Oparin, A. I.; Braunshteĭn, A. E.; Pasynskiĭ, A. G. (eds) *The Origin of Life on the Earth*. Pergamon. 1959. https://www.elsevier.com/books/the-origin-of-life-on-the-earth/oparin/978-1-4831-9737-1

13 Mesler, B.; Cleaves, H. J. *A Brief History of Creation: Science and the search for the origin of life*, page 180. 2016. W. W. Norton & Company, Inc.

14 Guerrero, R. 'Joan Oró (1923–2004)'. *International Microbiology*, volume 8, number 1, pages 63–68. 2005. http://revistes.iec.cat/index.php/IM/issue/view/786/showToc

15 Brack, A.; Ferris, J. P.; Lazcano, A.; Miller, S. L.; Schopf, J. W. 'In Memoriam: Professor Emeritus Joan "John" Oró (1923–2004)'. *Origins of Life and Evolution of Biospheres*, volume 35, issue 4, pages 297–298. 2005. https://doi.org/10.1007/s11084-005-3530-x

16 Oró, J. 'Synthesis of adenine from ammonium cyanide.' *Biochemical and Biophysical Research Communications*, volume 2, issue 6, pages 407–412. 1960. https://doi.org/10.1016/0006-291X(60)90138-8

17 Ferris, J. P.; Joshi, P. C.; Edelson, E. H.; Lawless, J. G. 'HCN: A plausible source of purines, pyrimidines and amino acids on the primitive Earth.' *Journal of Molecular Evolution*, volume 11, issue 4, pages 293–311. 1978. https://doi.org/10.1007/BF01733839

18 Oró, J. 'Studies in experimental organic cosmochemistry.' *Annals of the New York Academy of Sciences*, volume 108, issue 2, pages 464–481. 1963. https://doi.org/10.1111/j.1749-6632.1963.tb13402.x

19 Eaude, M. 'Obituary: Joan Oró.' *Guardian*, 9 September 2004. https://

www.theguardian.com/news/2004/sep/09/guardianobituaries.obituaries
20 Sullivan, W. 'Cyril Ponnamperuma, Scholar of Life's Origins, Is Dead at 71.' *New York Times*, 24 December 1994, page 10. https://www.nytimes.com/1994/12/24/obituaries/cyril-ponnamperuma-scholar-of-life-s-origins-is-dead-at-71.html
21 Navarro-González, R. 'In Memoriam: Cyril Andrew Ponnamperuma, 1923–1994.' *Origins of Life and Evolution of the Biosphere*, volume 28, issue 2, pages 105–108. 1998. https://doi.org/10.1023/A:1006532105251
22 Ponnamperuma, C.; Sagan, C.; Mariner, R. 'Synthesis of Adenosine Triphosphate under Possible Primitive Earth Conditions.' *Nature*, volume 199, pages 222–226. 1963. https://doi.org/10.1038/199222a0
23 Maruyama, K. 'The discovery of adenosine triphosphate and the establishment of its structure.' *Journal of the History of Biology*, volume 24, issue 1, pages 145–154. 1991. https://doi.org/10.1007/BF00130477
24 Sagan, C. *Cosmos*. 1980. Macdonald Futura.
25 Spangenburg, R.; Moser, K. *Carl Sagan: A biography*. 2004. Greenwood Publishing Group. https://www.penguinrandomhouse.com/books/231212/carl-sagan-by-ray-spangenburg-and-kit-moser/9781591026587/
26 Brahic, C. 'Volcanic lightning may have sparked life on Earth.' *New Scientist*, 16 October 2008. https://www.newscientist.com/article/dn14966-volcanic-lightning-may-have-sparked-life-on-earth/
27 Johnson, A. P.; Cleaves, H. J.; Dworkin, J. P.; Glavin, D. P.; Lazcano, A.; Bada, J. L. 'The Miller volcanic spark discharge experiment.' *Science*, volume 322, issue 5900, page 404. 2008. http://doi.org/10.1126/science.1161527
28 Parker, E. T.; Zhou, M.; Burton, A. S.; Glavin, D. P.; Dworkin, J. P.; Krishnamurthy, R.; Fernández, F. M.; Bada, J. L. 'A plausible simultaneous synthesis of amino acids and simple peptides on the primordial Earth.' *Angewandte Chemie*, volume 53, issue 31, pages 8132–8136. 2014. https://doi.org/10.1002/anie.201403683
29 Sherwood, E.; Oró, J. 'Cyanamide mediated syntheses under plausible primitive Earth conditions. Part I. The syntheses of p^1, p^2-dideoxythymidine 5'-pyrophosphate.' *Journal of Molecular Evolution*, volume 10, issue 3, pages 183–192. 1977. https://doi.org/10.1007/BF01764594
—Sherwood, E.; Joshi, A.; Oró, J. 'Cyanamide mediated syntheses under plausible primitive Earth conditions. II. The polymerization of deoxythymidine 5′–triphosphate.' *Journal of Molecular Evolution*, volume 10, issue 3, pages 193–209. 1977. https://doi.org/10.1007/BF01764595
—Nooner, D. W.; Sherwood, E.; More, M. A.; Oró, J. 'Cyanamide mediated syntheses under plausible primitive Earth conditions. III. Synthesis of peptides.' *Journal of Molecular Evolution*, volume 10, issue 3, pages 211–220. 1977. https://doi.org/10.1007/BF01764596
—Eichberg, J.; Sherwood, E.; Epps, D. E.; Oró, J. 'Cyanamide mediated

syntheses under plausible primitive Earth conditions. IV. The synthesis of acylglycerols.' *Journal of Molecular Evolution*, volume 10, issue 3, pages 221–230. 1977. https://doi.org/10.1007/BF01764597
—Epps, D. E.; Sherwood, E.; Eichberg, J.; Oró, J. 'Cyanamide mediated syntheses under plausible primitive Earth conditions. V. The synthesis of phosphatidic acids.' *Journal of Molecular Evolution*, volume 11, issue 4, pages 279–292. 1978. https://doi.org/10.1007/BF01733838

PART TWO: STRANGE OBJECTS

1 Monod, J. *Le hazard et la nécessité*. 1970. Éditions de Seuil, Paris. Published in English as *Chance and Necessity* by William Collins Sons & Co Ltd, 1972. Page 27.

Chapter 4: The DNA Revolution

1 Dahm, R. 'Discovering DNA: Friedrich Miescher and the early days of nucleic acid research.' *Human Genetics*, volume 122, issue 6, pages 565–581. 2008. https://doi.org/10.1007/s00439-007-0433-0
2 Miescher, F. 'Ueber die chemische Zusammensetzung der Eiterzellen.' *Medicinisch-chemische Untersuchungen*, volume 4, pages 441–460. 1871.
3 Jones, M. E. 'Albrecht Kossel, a biographical sketch.' *The Yale Journal of Biology and Medicine*, volume 26, issue 1, pages 80–97. 1953. https://www.ncbi.nlm.nih.gov/pmc/articles/PMC2599350/
4 Tipson, R. S. 'Obituary: Phoebus Aaron Theodor Levene, 1869–1940.' *Advances in Carbohydrate Chemistry*, volume 12, pages 1–12. 1957. https://doi.org/10.1016/S0096-5332(08)60202-7
5 Griffith, F. 'The significance of pneumococcal types.' *Journal of Hygiene*, volume 27, issue 2, pages 113–159. 1928. https://doi.org/10.1017/S0022172400031879
6 Avery, O. T.; MacLeod, C. M; McCarty, M. 'Studies on the chemical nature of the substance inducing transformation of pneumococcal types: Induction of transformation by a deoxyribonucleic acid fraction isolated from Pneumococcus Type III.' *Journal of Experimental Medicine*, volume 79, issue 2, pages 137–158. 1944. https://dx.doi.org/10.1084/jem.79.2.137
7 Hershey, A. D.; Chase, M. 'Independent functions of viral protein and nucleic acid in growth of bacteriophage.' *Journal of General Physiology*, volume 36, number 1, pages 39–56. 1952. https://doi.org/10.1085/jgp.36.1.39
8 Cobb, M. *Life's Greatest Secret*. 2015. Profile Books.
9 Sayre, A. *Rosalind Franklin and DNA*. 1975. W. W. Norton & Company.

10 Maddox, B. *Rosalind Franklin: The Dark Lady of DNA*. 2002. HarperCollins.
11 Elkin, L. 'Rosalind Franklin and the double helix.' *Physics Today*, volume 56, issue 3, page 42. 2003. https://doi.org/10.1063/1.1570771
12 Rich, A.; Stevens, C. F. 'Obituary: Francis Crick (1916–2004).' *Nature*, volume 430, issue 7002, pages 845–847. 2004. https://doi.org/10.1038/430845a
13 Watson, J. D.; Crick, F. H. C. 'Molecular structure of nucleic acids: A structure for deoxyribose nucleic acid.' *Nature*, volume 171, pages 737–738. 1953. https://doi.org/10.1038/171737a0
14 Wilkins, M. H. F.; Stokes, A. R.; Wilson, H. R. 'Molecular structure of nucleic acids: Molecular structure of deoxypentose nucleic acids.' *Nature*, volume 171, pages 738–740. 1953. https://doi.org/10.1038/171738a0
—Franklin, R. E.; Gosling, R. G. 'Molecular configuration in sodium thymonucleate.' *Nature*, volume 171, pages 740–741. 1953. https://doi.org/10.1038/171740a0
15 https://www.nobelprize.org/prizes/medicine/1962/summary/
16 Schrödinger, E. *What Is Life? The physical aspect of the living cell*. 1944. Cambridge University Press. http://www.whatislife.ie/downloads/What-is-Life.pdf
17 Soyfer, V. 'The consequences of political dictatorship for Russian science.' *Nature Reviews Genetics*, volume 2, issue 9, pages 723–729. 2001. https://doi.org/10.1038/35088598
18 Haldane, J. B. S. 'A physicist looks at genetics.' *Nature*, volume 355, pages 375–376. 1945. https://doi.org/10.1038/155375a0
19 Gamow, G. 'Possible relation between deoxyribonucleic acid and protein structures.' *Nature*, volume 173, page 318. 1954. https://doi.org/10.1038/173318a0
20 Palade, G. E. 'A small particulate component of the cytoplasm'. *Journal of Biophysical and Biochemical Cytology*, volume 1, issue 1, pages 59–68. 1955. https://doi.org/10.1083/jcb.1.1.59
21 Crick, F. H. C. 'On protein synthesis.' *The Symposia of the Society for Experimental Biology*, volume 12, pages 138–163. 1958. https://profiles.nlm.nih.gov/ps/access/scbbzy.pdf
22 Cobb, M. '60 years ago, Francis Crick changed the logic of biology.' *PLoS Biology*, 18 September 2017. https://doi.org/10.1371/journal.pbio.2003243
23 Hoagland, M. B.; Stephenson, M. L.; Scott, J. F.; Hecht, L. I.; Zamecnik, P. C. 'A soluble ribonucleic acid intermediate in protein synthesis.' *Journal of Biological Chemistry*, volume 231, issue 1, pages 241–257. 1958. https://www.ncbi.nlm.nih.gov/pubmed/13538965
24 Brenner, S.; Jacob, F.; Meselson, M. 'An Unstable Intermediate Carrying Information from Genes to Ribosomes for Protein Synthesis.' *Nature*,

volume 190, pages 576–581. 1961. http://doi.org/10.1038/190576a0
—Gros, F.; Hiatt, H.; Gilbert, W.; Kurland, C. G.; Risebrough, R.W.; Watson, J.D. 'Unstable ribonucleic acid revealed by pulse labelling of Escherichia coli.' *Nature*, volume 190, pages 581–585. 1961.

25 Nirenberg, M. W.; Matthaei, H. J. 'The Dependence of Cell-Free Protein Synthesis in E. coli upon Naturally Occurring or Synthetic Polyribonucleotides.' *Proceedings of the National Academy of Sciences of the United States of America*, volume 47, issue 10, pages 1588–1602. 1961. https://doi.org/10.1073/pnas.47.10.1588

26 Leder, P.; Nirenberg M. W. 'RNA codewords and protein synthesis, III: On the nucleotide sequence of a cysteine and a leucine RNA codeword.' *Proceedings of the National Academy of Sciences of the United States of America*, volume 52, issue 6, pages 1521–1529. 1964. https://doi.org/10.1073/pnas.52.6.1521

27 Fox, S. W. (ed) 'The Origins of Prebiological Systems and of their Molecular Matrices: Proceedings of a conference conducted at Wakulla Springs, Florida, on 27–30 October 1963 under the Auspices of the Institute for Space Biosciences, the Florida State University and the National Aeronautics and Space Administration', page 75. 1965. Elsevier, Inc.

Chapter 5: Crystal Clear

1 Henriques, M. 'The idea that life began as clay crystals is 50 years old.' BBC Earth, 24 August 2016. http://www.bbc.com/earth/story/20160823-the-idea-that-life-began-as-clay-crystals-is-50-years-old

2 https://www.argyll-bute.gov.uk/argyllcollection/artwork/landscape-fields-or-fieldscape

3 Cairns-Smith, A. G. 'The origin of life and the nature of the primitive gene.' *Journal of Theoretical Biology*, volume 10, issue 1, pages 53–88. 1966. https://doi.org/10.1016/0022-5193(66)90178-0

4 Dawkins, R. *The Blind Watchmaker*. 1986. Longman/Penguin.

5 Melius, P. 'Genetic Takeover and the Mineral Origins of Life.' *The Quarterly Review of Biology*, volume 59, number 1, page 65. 1984. https://doi.org/10.1086/413693

6 Weiss, A. 'Replication and evolution in inorganic systems.' *Angewandte Chemie International Edition*, volume 20, issue 10, pages 850–860. 1981. https://doi.org/10.1002/anie.198108501

7 Arrhenius, G.; Cairns-Smith, A. G.; Hartman, H.; Miller, S. L.; Orgel, L. E. 'Remarks on the Review Article "Replication and Evolution in Inorganic Systems" by Armin Weiss.' *Angewandte Chemie International Edition*, volume 25, issue 7, page 658. 1986. https://doi.org/10.1002/anie.198606581

8 Brack, A. 'Chapter 7.4: Clay minerals and the origin of life.' In: Bergaya,

F.; Theng, B. K. G.; Lagaly, G. (eds) *Handbook of Clay Science*. 2006. https://doi.org/10.1016/S1572-4352(05)01011-1
9 Lagaly, G.; Beneke, K. 'In memory of Armin Weiss, 1927–2010.' *Clays and Clay Minerals*, volume 59, number 1, pages 1–2. 2011. https://doi.org/10.1346/CCMN.2011.0590101
10 Cairns-Smith, A. G. *Seven Clues to the Origin of Life*. 1985. Cambridge University Press.
11 Ninio, J. 'A stir in the primeval soup.' *Nature*, volume 318, pages 119–120. 1985. https://doi.org/10.1038/318119b0
12 Fox, S. W.; Przybylski, A. '*Seven Clues to the Origin of Life: A Scientific Detective Story*. A. G. Cairns-Smith; *Origins: A Skeptic's Guide to the Creation of Life on Earth*. Robert Shapiro,' *The Quarterly Review of Biology*, volume 61, number 3, pages 397–398. 1986. https://doi.org/10.1086/415048
13 Cairns-Smith, A. G. *Evolving the Mind: On the nature of matter and the origin of consciousness*. 1996. Cambridge University Press.
—Cairns-Smith, A. G. *Secrets of the Mind: A tale of discovery and mistaken identity*. 1999. Copernicus.
14 Perceval-Maxwell, S. 'Evolving the Mind: On the nature of matter and the origin of consciousness (book).' *Mind, Culture and Activity*, volume 5, issue 4, pages 317–321. 1998. https://doi.org/10.1207/s15327884mca0504_7
—Dennett, D. C. 'Quantum incoherence.' *Nature*, volume 381, pages 485–486. 1996. https://doi.org/10.1038/381485a0
15 Peierls, R. E. 'Wolfgang Ernst Pauli, 1900–1958.' *Biographical Memoirs of Fellows of the Royal Society*, volume 5, pages 174–192. 1960. http://dx.doi.org/10.1098/rsbm.1960.0014

Chapter 6: The Schism

1 Sagan, C. *Cosmos*, pages 30–31. 1980. Macdonald Futura.
2 Shapiro, R. *Origins: A skeptic's guide to the creation of life on Earth*. 1986. Summit Books.
3 Holland, H. D. 'The oxygenation of the atmosphere and oceans.' *Philosophical Transactions of the Royal Society: Biological Sciences*, volume 361, issue 1470, pages 903–915. 2006. https://doi.org/10.1098/rstb.2006.1838
4 Fox, S. W. (ed) 'The Origins of Prebiological Systems and of their Molecular Matrices: Proceedings of a conference conducted at Wakulla Springs, Florida, on 27–30 October 1963 under the Auspices of the Institute for Space Biosciences, the Florida State University and the National Aeronautics and Space Administration', page 211. 1965. Elsevier, Inc.
5 Fox (1965), page 117.

6 Interview with Jim Kasting.
7 Subsequently published as: Bernal, J. D. 'The physical basis of life.' *Proceedings of the Physical Society. Section B*, volume 62, issue 10, pages 597–618. 1949. https://doi.org/10.1088/0370-1301/62/10/301
8 Ernst, W. G. 'William Rubey 1898–1974.' *National Academy of Sciences Biographical Memoirs*. 1978. Washington, DC. http://www.nasonline.org/publications/biographical-memoirs/memoir-pdfs/rubey-william.pdf
9 Rubey, W. W. 'Development of the hydrosphere and atmosphere with special reference to probable composition of the early atmosphere.' *Geological Society of America Special Paper*, volume 62, pages 631–650. 1955. https://doi.org/10.1130/SPE62-p631
10 https://aas.org/obituaries/donald-m-hunten-1925-2010
11 Hunten, D. M. 'The escape of light gases from planetary atmospheres.' *Journal of the Atmospheric Sciences*, volume 30, pages 1481–1494. 1973. https://doi.org/10.1175/1520-0469
—Walker, J. C. G. *Evolution of the Atmosphere*. 1977. Macmillan. https://catalogue.nla.gov.au/Record/784391
12 Kasting, J. F. 'Earth's early atmosphere.' *Science*, volume 259, issue 5097, pages 920–926. 1993. https://doi.org/10.1126/science.11536547
13 Zahnle, K.; Schaefer, L.; Fegley, B. 'Earth's earliest atmospheres.' *Cold Spring Harbor Perspectives in Biology*, volume 2, issue 10, a004895. 2010. https://dx.doi.org/10.1101/cshperspect.a004895
—Shaw, G. H. 'Earth's atmosphere – Hadean to early Proterozoic.' *Chemie der Erde – Geochemistry*, volume 68, issue 3, pages 235–264. 2008. https://doi.org/10.1016/j.chemer.2008.05.001
14 Kasting, J. F. 'Atmospheric composition of Hadean-early Archean Earth: The importance of CO.' *Geological Society of America Special Papers*, volume 504, pages 19–28. 2014. https://doi.org/10.1130/2014.2504(04)
15 Cleaves, H. J.; Chalmers, J. H.; Lazcano, A.; Miller, S. L.; Bada, J. L. 'A reassessment of prebiotic organic synthesis in neutral planetary atmospheres.' *Origins of Life and Evolution of Biospheres*, volume 38, issue 2, pages 105–115. 2008. http://doi.org/10.1007/s11084-007-9120-3
16 Shapiro (1986).
17 Mossel, E.; Steel, M. 'Random biochemical networks and the probability of self-sustaining autocatalysis.' *Journal of Theoretical Biology*, volume 233, issue 3, pages 327–336. 2005. https://doi.org/10.1016/j.jtbi.2004.10.011
18 Kauffman, S. A. 'Cellular homeostasis, epigenesis and replication in randomly aggregated macromolecular systems.' *Journal of Cybernetics*, volume 1, pages 71–96. 1971. https://doi.org/10.1080/01969727108545830
19 Kauffman, S. A. 'Autocatalytic sets of proteins.' *Journal of Theoretical Biology*, volume 119, issue 1, pages 1–24. 1986. https://doi.org/10.1016/S0022-5193(86)80047-9

20 Farmer, J. D.; Kauffman, S. A.; Packard, N. H. 'Autocatalytic replication of polymers.' *Physica D*, volume 22, issues 1–3, pages 50–67. 1986. https://doi.org/10.1016/0167-2789(86)90233-2
—Kauffman, S. A. *The Origins of Order.* 1993. Oxford University Press.

21 Eigen, M. 'Selforganization of matter and the evolution of biological macromolecules.' *Naturwissenschaften*, volume 58, issue 10, pages 465–523. 1971. https://doi.org/10.1007/BF00623322
—Eigen, M.; Schuster, P. 'The Hypercycle: A principle of natural selection. Part A: Emergence of the hypercycle.' *Naturwissenschaften*, volume 64, issue 11, pages 541–565. 1977. https://doi.org/10.1007/BF00450633
—Eigen, M.; Schuster, P. 'The Hypercycle: A principle of natural selection. Part B: The abstract hypercycle.' *Naturwissenschaften*, volume 65, issue 1, pages 7–41. 1978. https://doi.org/10.1007/BF00420631
—Eigen, M.; Schuster, P. 'The Hypercycle: A principle of natural selection. Part C: The realistic hypercycle.' *Naturwissenschaften*, volume 65, issue 7, pages 341–369. 1978. https://doi.org/10.1007/BF00439699

22 https://www.nobelprize.org/prizes/chemistry/1967/summary/

23 Hordijk, W.; Steel, M. 'Autocatalytic networks at the basis of life's origin and organization.' *Life*, volume 8, issue 4, pages 62–73. 2018. http://doi.org/10.3390/life8040062

24 Fox, S. W. (ed) 'The Origins of Prebiological Systems and of their Molecular Matrices: Proceedings of a conference conducted at Wakulla Springs, Florida, on 27–30 October 1963 under the Auspices of the Institute for Space Biosciences, the Florida State University and the National Aeronautics and Space Administration', page 216. 1965. Elsevier, Inc.

25 Hansma, H. G. 'Possible origin of life between mica sheets: does life imitate mica?' *Journal of Biomolecular Structure and Dynamics*, volume 31, issue 8, pages 888–895. 2013. https://doi.org/10.1080/07391102.2012.718528

26 Gold, T. 'The deep, hot biosphere.' *Proceedings of the National Academy of Sciences*, volume 89, issue 13, pages 6045–6049. 1992. https://doi.org/10.1073/pnas.89.13.6045

27 Ebisuzaki, T.; Maruyama, S. 'Nuclear geyser model of the origin of life: Driving force to promote the synthesis of building blocks of life.' *Geoscience Frontiers*, volume 8, issue 2, pages 275–298. 2017. https://doi.org/10.1016/j.gsf.2016.09.005

28 Benner, S. A.; Kim, H.-J.; Yang, Z. 'Setting the Stage: The History, Chemistry, and Geobiology behind RNA.' *Cold Spring Harbor Perspectives in Biology*, volume 4, issue 1, a003541. 2012. https://doi.org/10.1101/cshperspect.a003541

29 Noller, H. 'Carl Woese (1928–2012).' *Nature*, volume 493, page 610. 2013. https://doi.org/10.1038/493610a

30 Woese, C. R.; Fox, G. E. 'Phylogenetic structure of the prokaryotic domain: The primary kingdoms.' *Proceedings of the National Academy of Sciences*, volume 74, issue 11, pages 5088–5090. 1977. https://doi.org/10.1073/pnas.74.11.5088

31 Sagan, L. 'On the origin of mitosing cells.' *Journal of Theoretical Biology*, volume 14, issue 3, pages 225–274. 1967. https://doi.org/10.1016/0022-5193(67)90079-3

32 Hollinger, M. 'Life from Elsewhere – Early History of the Maverick Theory of Panspermia.' *Sudhoffs Archiv*, volume 100, issue 2, pages 188–205. 2016. https://www.jstor.org/stable/24913787

33 Kamminga, H. 'Life from space – A history of panspermia.' *Vistas in Astronomy*, volume 26, part 2, pages 67–86. 1982. https://doi.org/10.1016/0083-6656(82)90001-0

34 Fleischfresser, S. 'Over our heads: A brief history of panspermia.' *Cosmos*, 24 April 2018. https://cosmosmagazine.com/biology/over-our-heads-a-brief-history-of-panspermia

35 Arrhenius, S. A. *Worlds in the Making*. 1908. Harper & Brothers Publishers. Available online at: https://archive.org/details/worldsinmakingev00arrhrich

36 Crick, F. H. C.; Orgel, L. E. 'Directed panspermia.' *Icarus*, volume 19, issue 3, pages 341–346. 1973. https://doi.org/10.1016/0019-1035(73)90110-3

37 Burbidge, G. 'Sir Fred Hoyle. 24 June 1915 – 20 August 2001 Elected FRS 1957.' *Biographical Memoirs of Fellows of the Royal Society*, volume 49, pages 213–247. 2003. https://doi.org/10.1098/rsbm.2003.0013

38 Sagan, C. *Cosmos*, page 233. 1980. Macdonald Futura.

39 Wickramasinghe, C. 'Panspermia according to Hoyle.' *Astrophysics and Space Science*, volume 285, issue 2, pages 535–538. 2003. https://doi.org/10.1023/A:1025437920710

40 Naish, D. 'Alien viruses and *Archaeopteryx* forgery.' *Tetrapod Zoology*, 27 March 2012. https://blogs.scientificamerican.com/tetrapod-zoology/alien-viruses-and-archaeopteryx-forgery/

41 Sullivan, W. 'Creation debate is not limited to Arkansas trial.' *New York Times*, 27 December 1981, page 48. https://www.nytimes.com/1981/12/27/us/creation-debate-is-not-limited-to-arkansas-trial.html

42 https://www.panspermia.org/chandra.htm

43 Wickramasinghe, N. C.; Wallis, J.; Wallis, D. H.; Samaranayake, A. 'Fossil diatoms in a new carbonaceous meteorite.' *Journal of Cosmology*, volume 21, number 37. 2013. http://journalofcosmology.com/JOC21/PolonnaruwaRRRR.pdf

44 Plait, P. 'No, Diatoms Have Not Been Found in a Meteorite.' *Slate*, 15 January 2013. https://slate.com/technology/2013/01/life-in-a-meteorite-claims-by-n-c-wickramasinghe-of-diatoms-in-a-meteorite-are-almost-certainly-wrong.html

45 Myers, P. Z. 'Diatoms ... iiiiin spaaaaaaaaaaaace!' *Pharyngula*, 16 January 2013. https://freethoughtblogs.com/pharyngula/2013/01/16/diatomsiiiiin-spaaaaaaaaaaaace/

46 Wainwright, M.; Rose, C. E.; Baker, A. J.; Briston, K. J.; Wickramasinghe, N. C. 'Isolation of a diatom frustule fragment from the lower stratosphere (22–27km) – Evidence for a cosmic origin.' *Journal of Cosmology*, volume 22, pages 10183–10188. 2013. http://journalofcosmology.com/JOC22/milton_diatom.pdf

47 Steele, E. J.; Al-Mufti, S.; August, K. A.; Chandrajith, R.; Coghlan, J. P.; Coulson, S. G.; Ghosh, S.; Gillman, M.; Gorczynski, R. M.; Klyce, B.; Louis, G.; Mahanama, K.; Oliver, K. R.; Padron, J.; Qu, J.; Schuster, J. A.; Smith, W. E.; Snyder, D. P.; Steele, J. A.; Stewart, B. J.; Temple, R.; Tokoro, G.; Tout, C. A.; Unzicker, A.; Wainwright, M.; Wallis, J.; Wallis, D. H.; Wallis, M. K.; Wetherall, J.; Wickramasinghe, D. T.; Wickramasinghe, J. T.; Wickramasinghe, N. C.; Liu, Y. 'Cause of Cambrian Explosion – Terrestrial or Cosmic?' *Progress in Biophysics and Molecular Biology*, volume 136, pages 3–23. 2018. https://doi.org/10.1016/j.pbiomolbio.2018.03.004

48 Marx, K. to Lavrov, P. 18 June 1875. London. Reprinted in: *Marx & Engels: Collected Works, Volume 445, Letters 1874–79*. 2010. Lawrence & Wishart. Page 78.

49 Jean-Pierre de Vera, J.-P.; Alawi, M.; Backhaus, T.; Baqué, T.; Billi, D.; Böttger, U.; Berger, T.; Bohmeier, M.; Cockell, C.; Demets, R.; Noetzel, R. d. l. T.; Edwards, H.; Elsaesser, A.; Fagliarone, C.; Fiedler, A.; Foing, B.; Foucher, F.; Fritz, J.; Hanke, F.; Herzog, T.; Horneck, G.; Hübers, H.-W.; Huwe, B.; Joshi, J.; Kozyrovska, N.; Kruchten, M.; Lasch, P.; Lee, N.; Leuko, S.; Leya, T.; Lorek, A.; Martínez-Frías, J.; Meessen, J.; Moritz, S.; Moeller, R.; Olsson-Francis, K.; Onofri, S.; Ott, S.; Pacelli, C.; Podolich, O.; Rabbow, E.; Reitz, G.; Rettberg, P.; Reva, O.; Rothschild, L.; Sancho, L. G.; Schulze-Makuch, D.; Selbmann, L.; Serrano, P.; Szewzyk, U.; Verseux, C.; Wadsworth, J.; Wagner, D.; Westall, F.; Wolter, D.; Zucconi, L. 'Limits of Life and the Habitability of Mars: The ESA Space Experiment BIOMEX on the ISS.' *Astrobiology*, volume 19, number 2, pages 145–157. 2019. https://doi.org/10.1089/ast.2018.1897

PART THREE: SCATTERED, DIVIDED, LEADERLESS

1 https://www.darwinproject.ac.uk/letter/DCP-LETT-3272.xml

Chapter 7: The Other Long Molecule

1 Reynolds, J. A.; Tanford, C. *Nature's Robots: A History of Proteins* (Oxford Paperbacks). 2003. Oxford University Press.

2 Grimaux, E. *Lavoisier*. 1888.
3 Mulder, G. J. 'Sur la composition de quelques substances animals.' *Bulletin des Sciences Physiques et Naturelles en Néerlande*, volume 1, pages 104–119. 1838.
4 Hartley, H. 'Origin of the word "protein".' *Nature*, volume 168, page 244. 1951. https://doi.org/10.1038/168244a0
5 Sanger, F. 'The terminal peptides of insulin.' *Biochemical Journal*, volume 45, issue 5, pages 563–574. 1949b.
—Sanger, F.; Tuppy, H. 'The amino-acid sequence in the phenylalanyl chain of insulin. 1. The identification of lower peptides from partial hydrolysates.' *Biochemical Journal*, volume 49, issue 4, pages 463–481. 1951a.
—Sanger, F.; Tuppy, H. 'The amino-acid sequence in the phenylalanyl chain of insulin. 2. The investigation of peptides from enzymic hydrolysates.' *Biochemical Journal*, volume 49, issue 4, pages 481–490. 1951b.
—Sanger, F.; Thompson, E.O.P. 'The amino-acid sequence in the glycyl chain of insulin. 1. The identification of lower peptides from partial hydrolysates.' *Biochemical Journal*, volume 53, issue 3, pages 353–366. 1953a. https://dx.doi.org/10.1042/bj0530353
—Sanger, F.; Thompson, E.O.P. 'The amino-acid sequence in the glycyl chain of insulin. 2. The investigation of peptides from enzymic hydrolysates.' *Biochemical Journal*, volume 53, issue 3, pages 366–374. 1953b. https://dx.doi.org/10.1042/bj0530366.
6 Kendrew, J. C.; Bodo, G.; Dintzis, H. M.; Parrish, R. G.; Wyckoff, H.; Phillips, D. C. 'A three-dimensional model of the myoglobin molecule obtained by X-ray analysis.' *Nature*, volume 181, pages 662–666. 1958. https://doi.org/10.1038/181662a0
7 Sumner, J. B. 'The isolation and crystallization of the enzyme urease.' *Journal of Biological Chemistry*, volume 69, pages 435–441. 1926. http://www.jbc.org/content/69/2/435.citation
8 Fischer, E. 'Einfluss der Configuration auf die Wirkung der Enzyme.' *Berichte der deutschen chemischen Gesellschaft*, volume 27, pages 2985–2993. 1894. https://doi.org/10.1002/cber.18940270364
9 Macallum, A. B. 'On the origin of life on the globe.' *Transactions of the Canadian Institute*, volume 8, pages 423–441. 1908.
10 Schwartz, A. W. 'Sidney W. Fox, 1912–1998.' *Origins of Life and Evolution of the Biosphere*, volume 29, issue 1, pages 1–3. 1999. https://doi.org/10.1023/A:1006508001786
11 Fox, S. W. 'Evolution of Protein Molecules and Thermal Synthesis of Biochemical Substances.' *American Scientist*, volume 44, page 347. 1956. https://www.jstor.org/stable/27826834
12 Fox, S. W. *The Emergence of Life: Darwinian evolution from the inside*. 1988. Basic Books.

13 Fox, S. W.; Harada, K. 'The thermal copolymerization of amino acids common to protein.' *Journal of the American Chemical Society*, volume 82, issue 14, pages 3745–3751. 1960. http://doi.org/10.1021/ja01499a069
14 Fox, S. W.; Harada, K. 'Thermal Copolymerization of Amino Acids to a Product Resembling Protein.' *Science*, volume 128, issue 3333, page 1214. 1958. http://doi.org/10.1126/science.128.3333.1214
15 Fox, S. W.; Harada, K.; Kendrick, J. 'Production of Spherules from Synthetic Proteinoid and Hot Water.' *Science*, volume 129, page 1221. 1959. https://doi.org/10.1126/science.129.3357.1221-a
16 Fox, S. W. 'How did life begin?' *Science*, volume 132, issue 3421, pages 200–208. 1960. https://doi.org/10.1126/science.132.3421.200
17 Fox (1988), pages 38–39.
18 Fox, S. W. (ed) 'The Origins of Prebiological Systems and of their Molecular Matrices: Proceedings of a conference conducted at Wakulla Springs, Florida, on 27–30 October 1963 under the Auspices of the Institute for Space Biosciences, the Florida State University and the National Aeronautics and Space Administration', page 18. 1965. Elsevier, Inc.
19 Fox (1965), pages 374–375.
20 Fox, S. W. 'A theory of macromolecular and cellular origins.' *Nature*, volume 205, pages 328–340. 1965. https://doi.org/10.1038/205328a0
21 Fox, S. W. 'Metabolic microspheres.' *Naturwissenschaften*, volume 67, issue 8, pages 378–383. 1980. https://doi.org/10.1007/BF00405480
22 Fox (1988), page 84.
23 Dyson, F. J. 'Origins of Life.' In: *Nishina Memorial Lectures. Lecture Notes in Physics*, volume 746, pages 71–98. 2008. Springer, Tokyo. https://doi.org/10.1007/978-4-431-77056-5_5
24 Lee, D. H.; Granja, J. R.; Martinez, J. A.; Severin, K.; Ghadiri, M. R. 'A Self-Replicating Peptide.' *Nature*, volume 382, pages 525–528. 1996. https://doi.org/10.1038/382525a0
25 Lee, D. L.; Severin, K.; Yokobayashi, Y.; Ghadiri, M. R. 'Emergence of symbiosis in peptide self-replication through a hypercyclic network.' *Nature*, volume 390, pages 591–594. 1997. https://doi.org/10.1038/37569
26 Schreiber, A.; Huber, M. C.; Schiller, S. M. 'A prebiotic protocell model based on dynamic protein membranes accommodating anabolic reactions.' *bioRxiv*, 6 November 2018. https://doi.org/10.1101/463356

Chapter 8: Rise of the Replicators

1 Gilbert, W. 'Origin of life: The RNA world.' *Nature*, volume 319, page 618. 1986. https://doi.org/10.1038/319618a0
2 Belozersky, A. N.; Spirin, A. S. 'A Correlation between the Compositions of Deoxyribonucleic and Ribonucleic Acids.' *Nature*, volume 182, pages

111–112. 1958. https://doi.org/10.1038/182111a0

3 Fox, S. W. (ed) 'The Origins of Prebiological Systems and of their Molecular Matrices: Proceedings of a conference conducted at Wakulla Springs, Florida, on 27–30 October 1963 under the Auspices of the Institute for Space Biosciences, the Florida State University and the National Aeronautics and Space Administration', page 216. 1965. Elsevier, Inc.

4 Roberton, M. P.; Joyce, G. F. 'The Origins of the RNA World.' *Cold Spring Harbor Perspectives in Biology*, volume 4, issue 5, a003608. 2012. https://dx.doi.org/10.1101/cshperspect.a003608

5 Dunitz, J. D.; Joyce, G. F. 'Leslie Eleazer Orgel. 12 January 1927 – 27 October 2007.' *Biographical Memoirs of Fellows of the Royal Society*, volume 59, pages 277–289. 2013. https://doi.org/10.1098/rsbm.2013.0002

6 Alan Schwartz interview.

7 Orgel, L. E. 'Evolution of the genetic apparatus.' *Journal of Molecular Biology*, volume 38, issue 3, pages 381–393. 1968. https://doi.org/10.1016/0022-2836(68)90393-8
—Crick, F. H. C. 'The origin of the genetic code.' *Journal of Molecular Biology*, volume 38, issue 3, pages 367–379. 1968. https://doi.org/10.1016/0022-2836(68)90392-6

8 Woese, C. R. *The Genetic Code: The molecular basis for genetic expression*. 1967. Harper & Row.

9 Lohrmann, R.; Orgel, L. E. 'Prebiotic activation processes.' *Nature*, volume 244, pages 418–420. 1973. https://doi.org/10.1038/244418a0

10 Lohrmann, R.; Orgel, L. E. 'Template-directed synthesis of high molecular weight polynucleotide analogues.' *Nature*, volume 261, pages 342–344. 1976. https://doi.org/10.1038/261342a0

11 Thomas R. Cech – Biographical. NobelPrize.org. Nobel Media AB 2019. https://www.nobelprize.org/prizes/chemistry/1989/cech/biographical/

12 Cech, T. R. 'Self-Splicing and Enzymatic Activity of an Intervening Sequence RNA from *Tetrahymena*.' From *Nobel Lectures, Chemistry 1981–1990*, Editor-in-Charge Tore Frängsmyr, Editor Bo G. Malmström. World Scientific Publishing Co., Singapore. 1992. https://www.nobelprize.org/uploads/2018/06/cech-lecture.pdf

13 Kruger, K.; Grabowski, P. J.; Zaug, A. J.; Sands, J.; Gottschling, D. E.; Cech, T. R. 'Self-splicing RNA: Autoexcision and autocyclization of the ribosomal RNA intervening sequence of *Tetrahymena*.' *Cell*, volume 31, issue 1, pages 147–157. 1982. https://doi.org/10.1016/0092-8674(82)90414-7

14 Guerrier-Takada, C.; Gardiner, K.; Marsh, T.; Pace, N.; Altman, S. 'The RNA moiety of ribonuclease P is the catalytic subunit of the enzyme.' *Cell*, volume 35, issue 3, pages 849–857. 1983. https://doi.org/10.1016/0092-8674(83)90117-4

15 Been, M. D.; Cech, T. R. 'RNA as an RNA polymerase: net elongation

of an RNA primer catalyzed by the *Tetrahymena* ribozyme.' *Science*, volume 239, issue 4846, pages 1412–1416. 1988. https://doi.org/10.1126/science.2450400

16 Doudna, J. A.; Szostak, J. W. 'RNA-catalysed synthesis of complementary-strand RNA.' *Nature*, volume 339, pages 519–522. 1989. https://doi.org/10.1038/339519a0

17 Bartel, D. P.; Szostak, J. W. 'Isolation of new ribozymes from a large pool of random sequences.' *Science*, volume 261, issue 5127, pages 1411–1418. 1993. https://doi.org/10.1126/science.7690155

18 Orgel, L. E. 'Molecular replication.' *Nature*, volume 358, pages 203–209. 1992. https://doi.org/10.1038/358203a0

19 Johnston, W. K.; Unrau, P. J.; Lawrence, M. S.; Glasner, M. E.; Bartel, D. P. 'RNA-Catalyzed RNA Polymerization: Accurate and General RNA-Templated Primer Extension.' *Science*, volume 292, issue 5520, pages 1319–1325. 2001. https://doi.org/10.1126/science.1060786

20 Wochner, A.; Attwater, J.; Coulson, A.; Holliger, P. 'Ribozyme-Catalyzed Transcription of an Active Ribozyme.' *Science*, volume 332, issue 6026, pages 209–212. 2011. http://dx.doi.org/10.1126/science.1200752

21 Fox, S. W. (ed) 'The Origins of Prebiological Systems and of their Molecular Matrices: Proceedings of a conference conducted at Wakulla Springs, Florida, on 27–30 October 1963 under the Auspices of the Institute for Space Biosciences, the Florida State University and the National Aeronautics and Space Administration', page 18. 1965. Elsevier, Inc.

22 Schwartz, A. W. 'James P. Ferris 1932–2016.' *Origins of Life and Evolution of Biospheres*, volume 47, issue 1, pages 1–2. 2017. https://doi.org/10.1007/s11084-016-9505-2

23 Ferris, J. P.; Ertem, G.; Agarwal, V. 'Mineral catalysis of the formation of dimers of 5'-AMP in aqueous solution: The possible role of montmorillonite clays in the prebiotic synthesis of RNA.' *Origins of Life and Evolution of the Biosphere*, volume 19, issue 2, pages 165–178. 1989. https://doi.org/10.1007/BF01808150

24 Ferris, J. P.; Hill, A. R. Jr; Liu, R.; Orgel, L. E. 'Synthesis of long prebiotic oligomers on mineral surfaces.' *Nature*, volume 381, pages 59–61. 1996. https://doi.org/10.1038/381059a0

25 Schwartz, A. W.; Orgel, L. E. 'Template-directed synthesis of novel, nucleic acid-like structures.' *Science*, volume 228, issue 4699, pages 585–587. 1985. https://doi.org/10.1126/science.228.4699.585

26 Joyce, G. F.; Schwartz, A. W.; Miller, S. L.; Orgel, L. E. 'The case for an ancestral genetic system involving simple analogues of the nucleotides.' *Proceedings of the National Academy of Sciences*, volume 84, issue 13, pages 4398–4402. 1987. https://doi.org/10.1073/pnas.84.13.4398

27 Achilles, T.; von Kiedrowski, G. 'A self-replicating system from *three*

starting materials.' *Angewandte Chemie International Edition*, volume 32, issue 8, pages 1198–1201. 1993. https://doi.org/10.1002/anie.199311981

28 Sievers, D.; von Kiedrowski, G. 'Self-replication of complementary nucleotide-based oligomers.' *Nature*, volume 369, pages 221–224. 1994. https://doi.org/10.1038/369221a0

29 Nielsen, P. E.; Egholm, M.; Berg, R. H.; Buchardt, O. 'Sequence-selective recognition of DNA by strand displacement with a thymine-substituted polyamide.' *Science*, volume 254, issue 5037, pages 1497–1500. 1991. https://doi.org/10.1126/science.1962210

30 Wittung, P.; Nielsen, P. E.; Buchardt, O.; Egholm, M.; Nordén, B. 'DNA-like Double Helix formed by Peptide Nucleic Acid.' *Nature*, volume 368, issue 6471, pages 561–563. 1994. https://doi.org/10.1038/368561a0

31 Miller, S. L. 'Peptide nucleic acids and prebiotic chemistry.' *Nature Structural Biology*, volume 4, issue 3, pages 167–169. 1997. http://dx.doi.org/10.1038/nsb0397-167

32 Nelson, K. E.; Levy, M.; Miller, S. L. 'Peptide nucleic acids rather than RNA may have been the first genetic molecule.' *Proceedings of the National Academy of Sciences*, volume 97, issue 8, pages 3868–3871. 2000. https://doi.org/10.1073/pnas.97.8.3868

33 Schöning, K.-U.; Scholz, P.; Guntha, S.; Wu, X.; Krishnamurthy, R.; Eschenmoser, A. 'Chemical Etiology of Nucleic Acid Structure: The α-Threofuranosyl-(3'→2') Oligonucleotide System.' *Science*, volume 290, issue 5495, pages 1347–1351. 2000. https://doi.org/10.1126/science.290.5495.1347

34 Yu, H.; Zhang, S.; Chaput, J. C. 'Darwinian evolution of an alternative genetic system provides support for TNA as an RNA progenitor.' *Nature Chemistry*, volume 4, pages 183–187. 2012. http://dx.doi.org/10.1038/nchem.1241

35 Yonath, A.; Muessig, J.; Tesche, B.; Lorenz, S.; Erdmann, V. A.; Wittmann, H. G. 'Crystallization of the large ribosomal subunit from B. stearothermophilus.' *Biochemistry International*, volume 1, pages 428–35. 1980.

36 Yonath, A.; Bartunik, H. D.; Bartels, K. S.; Wittmann, H. G. 'Some X-ray diffraction patterns from single crystals of the large ribosomal subunit from Bacillus stearothermophilus.' *Journal of Molecular Biology*, volume 177, issue 1, pages 201–206. 1984. https://doi.org/10.1016/0022-2836(84)90066-4

37 Ban, N.; Nissen, P.; Hansen, J.; Moore, P. B.; Steitz, T. A. 'The Complete Atomic Structure of the Large Ribosomal Subunit at 2.4 Å Resolution.' *Science*, volume 289, issue 5481, pages 905–920. 2000. https://doi.org/10.1126/science.289.5481.905

—Nissen, P.; Hansen, J.; Ban, N.; Moore, P. B.; Steitz, T. A. 'The Structural Basis of Ribosome Activity in Peptide Bond Synthesis.' *Science*, volume 289, issue 5481, pages 920–930. 2000. https://doi.org/10.1126/science.289.5481.920

38 Cech, T. R. 'The ribosome is a ribozyme.' *Science*, volume 289, issue 5481, pages 878–879. 2000. https://doi.org/10.1126/science.289.5481.878

39 Schluenzen, F.; Tocilj, A.; Zarivach, R.; Harms, J.; Gluehmann, M.; Janell, D.; Bashan, A.; Bartels, H.; Agmon, I.; Franceschi, F.; Yonath, A. 'Structure of Functionally Activated Small Ribosomal Subunit at 3.3 Å Resolution.' *Cell*, volume 102, issue 5, pages 615–623. 2000. https://doi.org/10.1016/S0092-8674(00)00084-2
—Wimberly, B. T.; Brodersen, D. E.; Clemons, W. J. Jr; Morgan-Warren, R. J.; Carter, A. P.; Vonrhein, C.; Hartsch, T.; Ramakrishnan, V. 'Structure of the 30S ribosomal subunit.' *Nature*, volume 407, pages 327–339. 2000. https://doi.org/10.1038/35030006

40 The Nobel Prize in Chemistry 2009. NobelPrize.org. Nobel Media AB 2019. Saturday, 23 February 2019. https://www.nobelprize.org/prizes/chemistry/2009/summary/ https://doi.org/10.1126/science.289.5481.920

41 Kim, D. E.; Joyce, G. F. 'Cross-catalytic replication of an RNA ligase ribozyme.' *Chemistry & Biology*, volume 11, issue 11, pages 1505–1512. 2004. https://doi.org/10.1016/j.chembiol.2004.08.021
—Lincoln, T. A.; Joyce, G. F. 'Self-sustained replication of an RNA enzyme.' *Science*, volume 323, issue 5918, pages 1229–1232. 2009. https://doi.org/10.1126/science.1167856

42 Bernhardt, H. S. 'The RNA world hypothesis: the worst theory of the early evolution of life (except for all the others).' *Biology Direct*, volume 7, pages 23–37. 2012. https://doi.org/10.1186/1745-6150-7-23

Chapter 9: The Blobs

1 Lombard, J. 'Once upon a time the cell membranes: 175 years of cell boundary research.' *Biology Direct*, volume 9, issue 32. 2014. https://dx.doi.org/10.1186/s13062-014-0032-7

2 Gorter, E.; Grendel, F. 'On bimolecular layers of lipoids on the chromocytes of the blood.' *Journal of Experimental Medicine*, volume 41, issue 4, pages 439–443. 1925. http://doi.org/10.1084/jem.41.4.439

3 Danielli, J. F.; Davson, H. 'A contribution to the theory of permeability of thin films.' *Journal of Cellular and Comparative Physiology*, volume 5, issue 4, pages 495–508. 1935. https://doi.org/10.1002/jcp.1030050409

4 Singer, S. J.; Nicolson, G. L. 'The fluid mosaic model of the structure of cell membranes.' *Science*, volume 175, issue 4023, pages 720–731. 1972. http://doi.org/10.1126/science.175.4023.720

5 http://www.urcflow.com/jeewanu/home.html

6 Bahadur, K. 'Photosynthesis of amino-acids from paraformaldehyde and

potassium nitrate.' *Nature*, volume 173, issue 4415, page 1141. 1954. https://doi.org/10.1038/1731141a0

7 Bahadur, K.; Ranganayaki, S.; Verma, H. C.; Srivastava, R. B.; Agarwal, K. M. L.; Pandey, R. S.; Saxena, I.; Malviya, A. N.; Kumar, V.; Perti, O. N.; Pathak, H. D. 'Preparation of Jeewanu units capable of growth, multiplication and metabolic activity.' *Vijnana Parishad Anusandhan Patrika*, volume 6, page 63. 1963.

8 Bahadur, K.; Ranganayaki, S. 'Synthesis of Jeewanu, the Units Capable of Growth, Multiplication and Metabolic Activity. I. Preparation of Units Capable of Growth and Division and Having Metabolic Activity.' *Zentralblatt für Bakteriologie, Parasitenkunde, Infektionskrankheiten und Hygiene*, volume 117, issue 11, pages 567–574. 1964.

—Bahadur, K.; Verma, H. C.; Srivastava, R. B.; Agrawal, K. M. L.; Pandey, R. S.; Saxena, I.; Malviya, A. N.; Kumar, V.; Perti, O. N.; Pathak, H. D. 'Synthesis of Jeewanu, the Units Capable of Growth, Multiplication and Metabolic Activity. II. Photochemical Preparation of Growing and Multiplying Units with Metabolic Activities.' *Zentralblatt für Bakteriologie, Parasitenkunde, Infektionskrankheiten und Hygiene*, volume 117, issue 11, pages 575–584. 1964.

—Bahadur, K. 'Synthesis of Jeewanu, the Units Capable of Growth, Multiplication and Metabolic Activity. III. Preparation of Microspheres Capable of Growth and Division by Budding and Having Metabolic Activity with Peptides Prepared Thermally.' *Zentralblatt für Bakteriologie, Parasitenkunde, Infektionskrankheiten und Hygiene*, volume 117, issue 11, pages 585–602. 1964.

—Bahadur, K. 'Conversion of Lifeless Matter into the Living System.' *Zentralblatt für Bakteriologie, Parasitenkunde, Infektionskrankheiten und Hygiene*, volume 118, issue 11, pages 671–694. 1964.

9 Caren, L. D.; Ponnamperuma, C. 'A review of some experiments on the synthesis of "jeewanu".' *NASA Technical Memorandum X-1439*. Moffett Field, California: Ames Research Center. https://ntrs.nasa.gov/archive/nasa/casi.ntrs.nasa.gov/19670026284.pdf

10 Grote, M. '*Jeewanu*, or the "particles of life".' *Journal of Biosciences*, volume 36, issue 4, pages 563–570. 2011. https://doi.org/10.1007/s12038-011-9087-0

11 Kumar, V. K.; Rai, R. K. 'Cytochemical characterisation of photochemically formed, self-sustaining, abiogenic, protocell-like, supramolecular assemblies "Jeewanu".' *International Journal of Life Sciences*, volume 6, issue 4, pages 877–884. 2018. http://www.ijlsci.in/abstract-6-4-5

12 Heap, B.; Gregoriadis, G. 'Alec Douglas Bangham. 10 November 1921 – 9 March 2010.' *Biographical Memoirs of Fellows of the Royal Society*, volume 57, pages 25–43. 2011. https://doi.org/10.1098/rsbm.2011.0004

13 Hargreaves, W. R.; Mulvihill, S. J.; Deamer, D. W. 'Synthesis of phospholipids and membranes in prebiotic conditions.' *Nature*, volume 266, issue 5597, pages 78–80. 1977. https://doi.org/10.1038/266078a0
14 Hargreaves, W. R.; Deamer, D. W. 'Liposomes from ionic, single-chain amphiphiles.' *Biochemistry*, volume 17, issue 18, pages 3759–3768. 1978. http://doi.org/10.1021/bi00611a014
15 Deamer, D. W.; Oró, J. 'Role of lipids in prebiotic structures.' *Biosystems*, volume 12, issues 3–4, pages 167–175. 1980. https://doi.org/10.1016/0303-2647(80)90014-3
16 Deamer, D. W.; Barchfeld, G. L. 'Encapsulation of macromolecules by lipid vesicles under simulated prebiotic conditions.' *Journal of Molecular Evolution*, volume 18, issue 3, pages 203–206. 1982. https://doi.org/10.1007/BF01733047
17 Stillwell, W. 'Facilitated diffusion as a method for selective accumulation of materials from the primordial oceans by a lipid-vesicle protocell.' *Origins of Life*, volume 10, issue 3, pages 277–292. 1980. https://doi.org/10.1007/BF00928406
—Yanagawa, H.; Ogawa, Y.; Kojima, K.; Ito, M. 'Construction of protocellular structures under simulated primitive earth conditions.' *Origins of Life and Evolution of the Biosphere*, volume 18, issue 3, pages 179–207. 1988. https://doi.org/10.1007/BF01804670
18 https://bettinaheinzart.com
19 Morowitz, H. J.; Heinz, B.; Deamer, D. W. 'The chemical logic of a minimum protocell.' *Origins of Life and Evolution of the Biosphere*, volume 18, issue 3, pages 281–287. 1988. https://doi.org/10.1007/BF01804674
20 Deamer, D. W. 'Polycyclic aromatic hydrocarbons: Primitive pigment systems in the prebiotic environment.' *Advances in Space Research*, volume 12, issue 4, pages 183–189. 1992. https://doi.org/10.1016/0273-1177(92)90171-S
21 Morowitz, H. J. *Beginnings of Cellular Life: Metabolism recapitulates biogenesis*. 1992. Yale University Press.
22 https://web.archive.org/web/20070611152326/http://www.cts.cuni.cz/conf98/luisi.htm
23 Luisi, P. L.; Houshmand, Z. *Mind and Life: Discussions with the Dalai Lama on the nature of reality*. 2008. Columbia University Press.
24 Luisi, P. L.; Varela, F. J. 'Self-replicating micelles – A chemical version of a minimal autopoietic system.' *Origins of Life and Evolution of the Biosphere*, volume 19, issue 6, pages 633–643. 1989. https://doi.org/10.1007/BF01808123
25 Bachmann, P. A.; Walde, P.; Luisi, P. L.; Lang, J. 'Self-replicating reverse micelles and chemical autopoiesis.' *Journal of the American Chemical Society*, volume 112, issue 22, pages 8200–8201. 1990. http://doi.org/10.1021/ja00178a073

26 Bachmann, P. A.; Walde, P.; Luisi, P. L.; Lang, J. 'Self-replicating micelles: aqueous micelles and enzymatically driven reactions in reverse micelles.' *Journal of the American Chemical Society*, volume 113, issue 22, pages 8204–8209. 1991. http://doi.org/10.1021/ja00022a002
27 Bachmann, P. A.; Luisi, P. L.; Lang, J. 'Autocatalytic self-replicating micelles as models for prebiotic structures.' *Nature*, volume 357, issue 6373, pages 57–59. 1992. https://doi.org/10.1038/357057a0
28 Walde, P.; Wick, R.; Fresta, M.; Mangone, A.; Luisi, P. L. 'Autopoietic Self-Reproduction of Fatty Acid Vesicles'. *Journal of the American Chemical Society*, volume 116, issue 26, pages 11649–11654. 1994. http://doi.org/10.1021/ja00105a004
29 Wick, R.; Luisi, P. L. 'Enzyme-containing liposomes can endogenously produce membrane-constituting lipids.' *Chemistry & Biology*, volume 3, issue 4, pages 277–285. 1996. https://doi.org/10.1016/S1074-5521(96)90107-6
30 Chakrabarti, A. C.; Breaker, R. R.; Joyce, G. F.; Deamer, D. W. 'Production of RNA by a polymerase protein encapsulated within phospholipid vesicles.' *Journal of Molecular Evolution*, volume 39, issue 6, pages 555–559. 1994. https://doi.org/10.1007/BF00160400
31 Oberholzer, T.; Nierhaus, K. H.; Luisi, P. L. 'Protein Expression in Liposomes.' *Biochemical and Biophysical Research Communications*, volume 261, issue 2, pages 238–241. 1999. https://doi.org/10.1006/bbrc.1999.0404
32 Luisi, P. L.; Walde, P.; Oberholzer, T. 'Lipid vesicles as possible intermediates in the origin of life.' *Current Opinion in Colloid & Interface Science*, volume 4, issue 1, pages 33–39. 1999. https://doi.org/10.1016/S1359-0294(99)00012-6
33 Segré, D.; Ben-Eli, D.; Deamer, D. W.; Lancet, D. 'The Lipid World.' *Origins of Life and Evolution of the Biosphere*, volume 31, issue 1–2, pages 119–145. 2001. https://doi.org/10.1023/A:1006746807104
34 Lancet, D.; Zidovetski, R.; Markovitch, O. 'Systems protobiology: origin of life in lipid catalytic networks.' *Journal of the Royal Society Interface*, volume 15, issue 144, article 20180159. 2018. https://doi.org/10.1098/rsif.2018.0159

Chapter 10: The Need for Power

1 Kragh, H. 'Photon: New light on an old name.' *arXiv*, 1401:0293. 2014. https://arxiv.org/abs/1401.0293
2 Howard, J. N. 'Profile in optics: Leonard Thompson Troland.' *Optics and Photonics News*, volume 19, issue 6, page 20. 2008. https://doi.org/10.1364/OPN.19.6.000020

3 Roback, A. A. 'Obituary: Leonard Thompson Troland.' *Science*, volume 76, issue 1958, pages 26–28. 1932. http://dx.doi.org/10.1126/science.76.1958.26
4 Troland, L. T. 'The chemical origin and regulation of life.' *The Monist*, volume 24, issue 1, pages 92–133. 1914. https://www.jstor.org/stable/27900476
5 Eddington, A. S. *The Nature of the Physical World*. 1928. Cambridge University Press. Available at: http://henry.pha.jhu.edu/Eddington.2008.pdf
6 Atkins, P. W. *The Second Law: Energy, chaos, and form*. 1984. Scientific American Library.
7 Eakin, R. E. 'An approach to the evolution of metabolism.' *Proceedings of the National Academy of Sciences*, volume 49, issue 3, pages 360–366. 1963. https://www.ncbi.nlm.nih.gov/pmc/articles/PMC299833/
8 Evans, M. C.; Buchanan, B. B.; Arnon, D. I. 'A new ferredoxin-dependent carbon reduction cycle in a photosynthetic bacterium.' *Proceedings of the National Academy of Sciences*, volume 55, issue 4, pages 928–934. 1966. https://doi.org/10.1073/pnas.55.4.928
9 Morowitz, H. J. *Energy Flow in Biology: Biological organization as a problem in thermal physics*. 1968. Academic Press, Inc.
10 Morowitz, H. J. 'Physical background of cycles in biological systems.' *Journal of Theoretical Biology*, volume 13, pages 60–62. 1966. https://doi.org/10.1016/0022-5193(66)90007-5
11 Kuhn, H. 'Self-organization of molecular systems and evolution of the genetic apparatus.' *Angewandte Chemie International Edition*, volume 11, issue 9, pages 798–820. 1972. https://doi.org/10.1002/anie.197207981
12 Phone interview with Günter Wächtershäuser.
13 Wächtershäuser, G. 'The case for the chemoautotrophic origin of life in an iron-sulfur world.' *Origins of Life and Evolution of the Biosphere*, volume 20, issue 2, pages 173–176. 1990. https://doi.org/10.1007/BF01808279
14 Wächtershäuser, G. 'Pyrite Formation, the First Energy Source for Life: a Hypothesis.' *Systematic and Applied Microbiology*, volume 10, issue 3, pages 207–210. 1988. https://doi.org/10.1016/S0723-2020(88)80001-8
—Wächtershäuser, G. 'Before enzymes and templates: theory of surface metabolism.' *Microbiology Reviews*, volume 52, issue 4, pages 452–484. 1988. http://www.ncbi.nlm.nih.gov/pubmed/3070320
15 Stetter, K. O.; König, H.; Stackebrandt, E. 'Pyrodictium gen. nov., a New Genus of Submarine Disc-Shaped Sulphur Reducing Archaebacteria Growing Optimally at 105°C.' *Systematic and Applied Microbiology*, volume 4, issue 4, pages 535–551. 1983. https://doi.org/10.1016/S0723-2020(83)80011-3
16 https://extinctmonsters.net/2013/07/17/the-osborn-problem/

17 Osborn, H. F. *The Origin and Evolution of Life: On the theory of action, reaction and interaction of energy*. 1916. The Science Press for the United States of America. Available at: https://archive.org/details/in.ernet.dli.2015.42611/page/n5
18 Wächtershäuser, G. 'Evolution of the first metabolic cycles.' *Proceedings of the National Academy of Sciences*, volume 87, issue 1, pages 200–204. 1990. https://doi.org/10.1073/pnas.87.1.200
19 De Duve, C.; Miller, S. L. 'Two-dimensional life?' *Proceedings of the National Academy of Sciences*, volume 88, issue 22, pages 10014–10017. 1991. https://doi.org/10.1073/pnas.88.22.10014
20 Wächtershäuser, G. 'Groundworks for an evolutionary biochemistry: The iron-sulfur world.' *Progress in Biophysics and Molecular Biology*, volume 58, issue 2, pages 85–201. 1992. https://doi.org/10.1016/0079-6107(92)90022-X
21 Drobner, E.; Huber, H.; Wächtershäuser, G.; Rose, D.; Stetter, K. O. 'Pyrite formation linked with hydrogen evolution under anaerobic conditions.' *Nature*, volume 346, pages 742–744. 1990. https://doi.org/10.1038/346742a0
22 Hafenbradl, D.; Keller, M.; Wächtershäuser, G.; Stetter, K. O. 'Primordial amino acids by reductive amination of α-oxo acids in conjunction with the oxidative formation of pyrite.' *Tetrahedron Letters*, volume 36, issue 29, pages 5179–5182. 1995. https://doi.org/10.1016/0040-4039(95)01008-6
23 Huber, C.; Wächtershäuser, G. 'Activated Acetic Acid by Carbon Fixation on (Fe,Ni)S Under Primordial Conditions.' *Science*, volume 276, issue 5310, pages 245–247. 1997. https://doi.org/10.1126/science.276.5310.245
24 De Duve, C. *Blueprint for a Cell: The nature and origin of life*. 1991. Burlington, NC: Carolina Biological Supply Company (Neil Patterson Publisher).
25 Huber, C.; Wächtershäuser, G. 'Peptides by Activation of Amino Acids with CO on (Ni,Fe)S Surfaces: Implications for the Origin of Life.' *Science*, volume 281, issue 5377, pages 670–672. 1998. https://doi.org/10.1126/science.281.5377.670
26 Huber, C.; Eisenreich, W.; Hecht, S.; Wächtershäuser, G. 'A possible primordial peptide cycle.' *Science*, volume 301, issue 5635, pages 938–940. 2003. https://doi.org/10.1126/science.1086501

Chapter 11: Born in the Depths

1 Crane, K.; Normark, W. R. 'Hydrothermal activity and crestal structure of the East Pacific Rise at 21°N.' *Journal of Geophysical Research*, volume 82, issue 33, pages 5336–5348. 1977. https://doi.org/10.1029/jb082i033p05336
2 https://www.whoi.edu/feature/history-hydrothermal-vents/discovery/1977.html
3 Ballard, R. D. 'Notes on a major oceanographic find.' *Oceanus*, volume 20,

number 3, pages 35–44. 1977. Available at: https://divediscover.whoi.edu/archives/ventcd/pdf/BallardOc77Notes.pdf

4 Lonsdale, P. 'Clustering of suspension-feeding macrobenthos near abyssal hydrothermal vents at oceanic spreading centers.' *Deep-Sea Research*, volume 24, issue 9, pages 857–863. 1977. https://doi.org/10.1016/0146-6291(77)90478-7

—Corliss, J. B.; Dymond, J.; Gordon, L. I.; Edmond, J. M.; von Herzen, R. P.; Ballard, R. D.; Green, K.; Williams, D.; Bainbridge, A.; Crane, K.; van Andel, T. H. 'Submarine Thermal Springs on the Galápagos Rift.' *Science*, volume 203, issue 4385, pages 1073–1083. 1979. http://doi.org/10.1126/science.203.4385.1073

5 Spiess, F. N.; Macdonald, K. C.; Atwater, T.; Ballard, R.; Carranza, A.; Cordoba, D.; Cox, C.; Diaz Garcia, V. M.; Francheteau, J.; Guerrero, J.; Hawkins, J.; Hessler, R.; Haymon R.; Juteau, T.; Kastner, M.; Larson, R.; Luyendyk, B.; Macdougall, J. D.; Miller, S.; Normark, W.; Orcutt, J.; Rangin, C. 'Hot springs and geophysical experiments on the East Pacific Rise.' *Science*, volume 207, issue 4438, pages 1421–1444. 1980. https://doi.org/10.1126/science.207.4438.1421

6 Cavanaugh, C. M.; Gardiner, S. L.; Jones, M. L.; Jannasch, H. W.; Waterbury, J. B. 'Prokaryotic Cells in the Hydrothermal Vent Tube Worm *Riftia pachyptila* Jones: Possible Chemoautotrophic Symbionts.' *Science*, volume 213, issue 4505, pages 340–342. 1981. https://doi.org/10.1126/science.213.4505.340

7 Corliss, J. B.; Baross, J. A.; Hoffman, S. E. 'An hypothesis concerning the relationship between submarine hot springs and the origin of life on Earth.' *Oceanologica Acta*, volume 4 (special supplement), pages 59–69. 1981. https://archimer.ifremer.fr/doc/00245/35661/

—Baross, J. A.; Hoffman, S. E. 'Submarine hydrothermal vents and associated gradient environments as sites for the origin and evolution of life.' *Origins of Life and Evolution of the Biosphere*, volume 15, issue 4, pages 327–345. 1985. https://doi.org/10.1007/BF01808177

8 Miller, S. L.; Bada, J. L. 'Submarine hot springs and the origin of life.' *Nature*, volume 334, issue 6183, pages 609–611. 1988. http://dx.doi.org/10.1038/334609a0

9 Whitfield, J. 'Origin of life: Nascence man.' *Nature*, volume 459, issue 7245, pages 316–319. 2009. https://doi.org/10.1038/459316a

10 Larter, R. C. L.; Boyce, A. J.; Russell, M. J. 'Hydrothermal pyrite chimneys from the Ballynoe baryte deposit, Silvermines, County Tipperary, Ireland.' *Mineralium Deposita*, volume 16, issue 2, pages 309–317. 1981. https://doi.org/10.1007/BF00202742

—Boyce, A. J.; Coleman, M. L.; Russell, M. J. 'Formation of fossil hydrothermal chimneys and mounds from Silvermines, Ireland.'

Nature, volume 306, issue 5943, pages 545–550. 1983. https://doi.org/10.1038/306545a0

11 Russell, M. J.; Hall, A. J.; Cairns-Smith, A. G.; Braterman, P. S. 'Submarine hot springs and the origin of life.' *Nature*, volume 336, issue 6195, page 117. 1988. https://doi.org/10.1038/336117a0

12 Russell, M. J.; Hall, A. J.; Gize, A. P. 'Pyrite and the origin of life.' *Nature*, volume 344, issue 6265, page 387. 1990. https://doi.org/10.1038/344387b0

13 Leduc, S. *The Mechanism of Life*. 1911. Redman Company, New York. https://www.gutenberg.org/files/33862/33862-h/33862-h.htm

14 Russell, M. J.; Hall, A. J.; Turner, D. '*In vitro* growth of iron sulphide chimneys: possible culture chambers for origin-of-life experiments.' *Terra Nova*, volume 1, issue 3, pages 238–241. 1989. https://doi.org/10.1111/j.1365-3121.1989.tb00364.x

15 Prebble, J.; Weber, B. *Wandering in the Gardens of the Mind: Peter Mitchell and the making of Glynn*. 2003. Oxford University Press.

16 Slater, E. C. 'Peter Dennis Mitchell. 29 September 1920 – 10 April 1992.' *Biographical Memoirs of Fellows of the Royal Society*, volume 40, pages 282–305. 1994. https://doi.org/10.1098/rsbm.1994.0040

17 Cockell, C. *The Equations of Life: How physics shapes evolution*. 2018. Basic Books. Page 164.

18 Mitchell, P. 'Coupling of phosphorylation to electron and hydrogen transfer by a chemiosmotic type of mechanism.' *Nature*, volume 191, issue 4784, pages 144–148. 1961. https://doi.org/10.1038/191144a0

19 Mitchell, P.; Moyle, J. 'Chemiosmotic hypothesis of oxidative phosphorylation.' *Nature*, volume 213, issue 5072, pages 137–139. 1967. https://doi.org/10.1038/213137a0

20 https://www.nobelprize.org/prizes/chemistry/1978/summary/

21 Mulkidjanian, A. Y.; Galperin, M. Y.; Koonin, E. V. 'Co-evolution of primordial membranes and membrane proteins.' *Trends in Biochemical Sciences*, volume 34, issue 4, pages 206–215. 2009. https://doi.org/10.1016/j.tibs.2009.01.005

22 Russell, M. J.; Daniel, R. M.; Hall, A. J. 'On the emergence of life via catalytic iron-sulphide membranes.' *Terra Nova*, volume 5, issue 4, pages 343–347. 1993. https://doi.org/10.1111/j.1365-3121.1993.tb00267.x

23 Cairns-Smith, A. G.; Hall, A. J.; Russell, M. J. 'Mineral Theories of the Origin of Life and an Iron Sulfide Example.' In: Holm, N. G. (ed) *Marine Hydrothermal Systems and the Origin of Life*, pages 161–180. 1992. Springer, Dordrecht. https://doi.org/10.1007/978-94-011-2741-7_9

—Macleod, G.; McKeown, C.; Hall, A. J.; Russell, M. J. 'Hydrothermal and oceanic pH conditions of possible relevance to the origin of life.' *Origins of Life and Evolution of the Biosphere*, volume 24, issue 1, pages 19–41. 1994. https://doi.org/10.1007/BF01582037

—Kaschke, M.; Russell, M. J.; Cole, W. J. '[FeS/FeS_2]. A redox system for the origin of life (some experiments on the pyrite-hypothesis).' *Origins of Life and Evolution of the Biosphere*, volume 24, issue 1, pages 43–56. 1994. https://doi.org/10.1007/BF01582038

—Russell, M. J.; Daniel, R. M.; Hall, A. J.; Sherringham, J. A. 'A hydrothermally precipitated catalytic iron sulphide membrane as a first step toward life.' *Journal of Molecular Evolution*, volume 39, issue 3, pages 231–243. 1994. https://doi.org/10.1007/BF00160147

—Russell, M. J.; Hall, A. J. 'The emergence of life from iron monosulphide bubbles at a submarine hydrothermal redox and pH front.' *Journal of the Geological Society*, volume 154, issue 3, pages 377–402. 1997. https://doi.org/10.1144/gsjgs.154.3.0377

24 Kelley, D. S. 'From the Mantle to Microbes: The Lost City hydrothermal field.' *Oceanography*, volume 18, number 3, pages 32–45. 2005. http://dx.doi.org/10.5670/oceanog.2005.23

25 Kelley, D. S.; Karson, J. A.; Blackman, D. K.; Früh-Green, G. L.; Butterfield, D. A.; Lilley, M. D.; Olson, E. J.; Schrenk, M. O.; Roe, K. K.; Lebon, G. T.; Rivizzigno, P.; the AT3-60 Shipboard Party. 'An off-axis hydrothermal vent field near the Mid-Atlantic Ridge at 30° N.' *Nature*, volume 412, issue 6843, pages 145–149. 2001. https://doi.org/10.1038/35084000

26 Kelley, D. S.; Karson, J. A.; Früh-Green, G. L.; Yoerger, D. R.; Shank, T. M.; Butterfield, D. A.; Hayes, J. M.; Schrenk, M. O.; Olson, E. J.; Proskurowski, G.; Jakuba, M.; Bradley, A.; Larson, B.; Ludwig, K.; Glickson, D.; Buckman, K.; Bradley, A. S.; Brazelton, W. J.; Roe, K.; Elend, M. J.; Delacour, A.; Bernasconi, S. M.; Lilley, M. D.; Baross, J. A.; Summons, R. E.; Sylva, S. P. 'A Serpentinite-Hosted Ecosystem: The Lost City Hydrothermal Field.' *Science*, volume 307, issue 5714, pages 1428–1434. 2005. https://doi.org/10.1126/science.1102556

27 Baaske, P.; Weinert, F. M.; Duhr, S.; Lemke, K. H.; Russell, M. J.; Braun, D. 'Extreme accumulation of nucleotides in simulated hydrothermal pore systems.' *Proceedings of the National Academy of Sciences*, volume 104, issue 22, pages 9346–9351. 2007. https://doi.org/10.1073/pnas.0609592104

28 Martin, W.; Russell, M. J. 'On the origins of cells: a hypothesis for the evolutionary transitions from abiotic geochemistry to chemoautotrophic prokaryotes, and from prokaryotes to nucleated cells.' *Philosophical Transactions of the Royal Society B: Biological Sciences*, volume 358, issue 1429, pages 59–83. 2003. https://doi.org/10.1098/rstb.2002.1183

—Russell, M. J.; Martin, W. 'The rocky roots of the acetyl-CoA pathway.' *Trends in Biochemical Sciences*, volume 29, issue 7, pages 358–363. 2004. https://doi.org/10.1016/j.tibs.2004.05.007

—Martin, W.; Russell, M. J. 'On the origin of biochemistry at an alkaline hydrothermal vent.' *Philosophical Transactions of the Royal Society*

B: Biological Sciences, volume 362, issue 1486, pages 1887–1925. 2006. https://doi.org/10.1098/rstb.2006.1881

29 Lane, N. *The Vital Question: Why is life the way it is?* 2015. Profile Books Ltd.

30 Herschy, B.; Whicher, A.; Camprubi, E.; Watson, C.; Dartnell, L.; Ward, J.; Evans, J. R. G.; Lane, N. 'An Origin-of-Life Reactor to Simulate Alkaline Hydrothermal Vents.' *Journal of Molecular Evolution*, volume 79, issue 5–6, pages 213–227. 2014. https://doi.org/10.1007/s00239-014-9658-4

31 Wächtershäuser, G. 'In praise of error.' *Journal of Molecular Evolution*, volume 82, issue 2–3, pages 75–80. 2016. https://doi.org/10.1007/s00239-015-9727-3

32 Barge, L. M.; Flores, E.; Baum, M. M.; VanderVelde D. G.; Russell, M. J. 'Redox and pH gradients drive amino acid synthesis in iron oxyhydroxide mineral systems.' *Proceedings of the National Academy of Sciences*, volume 116, issue 11, pages 4828–4833. 2019. https://doi.org/10.1073/pnas.1812098116

33 Lombard, J.; López-García, P.; Moreira, D. 'The early evolution of lipid membranes and the three domains of life.' *Nature Reviews Microbiology*, volume 10, pages 507–515. 2012. https://doi.org/10.1038/nrmicro2815

34 Sutherland, J. D. 'Opinion: Studies on the origin of life – the end of the beginning.' *Nature Reviews Chemistry*, volume 1, article 0012. 2017. https://doi.org/10.1038/s41570-016-0012

35 Jackson, J. B. 'Natural pH gradients in hydrothermal alkali vents were unlikely to have played a role in the origin of life.' *Journal of Molecular Evolution*, volume 83, issue 1–2, pages 1–11. 2016. https://doi.org/10.1007/s00239-016-9756-6

36 Morowitz, H. J. *Beginnings of Cellular Life: Metabolism recapitulates biogenesis*, page 143. 1992. Yale University Press.

37 Russell, M. J. 'Green Rust: The Simple Organizing "Seed" of All Life?' *Life*, volume 8, issue 3, article E35. 2018. https://doi.org/10.3390/life8030035

38 Weiss, M. C.; Sousa, F. L.; Mrnjavac, N.; Neukirchen, S.; Roettger, M.; Nelson-Sathi, S.; Martin, W. F. 'The physiology and habitat of the last universal common ancestor.' *Nature Microbiology*, volume 1, issue 9, article 16116. 2016. https://doi.org/10.1038/nmicrobiol.2016.116

PART FOUR: REUNIFICATION

Chapter 12: Mirrors

1 Gounelle, M. 'The meteorite fall at L'Aigle and the Biot report: exploring the cradle of meteoritics.' In: *The History of Meteoritics and Key Meteoritic Collections: Fireballs, Falls and Finds*, Geological Society,

London, Special Publications, volume 256, pages 73–89. 2006. https://doi.org/10.1144/GSL.SP.2006.256.01.03

2 Biot, J. B. 'Phénomènes de polarisation successive, observés dans des fluides homogènes.' ('Phenomenon of successive polarization, observed in homogeneous fluids.') *Bulletin des Sciences, par la Société Philomatique de Paris*, pages 190–192. 1815. https://www.biodiversitylibrary.org/item/26553#page/196/mode/1up (in French)

3 Pasteur, L. 'Mémoire sur la relation qui peut exister entre la forme cristalline et la composition chimique, et sur la cause de la polarisation rotatoire.' ('Memoir on the relationship that can exist between crystalline form and chemical composition, and on the cause of rotary polarization.') *Comptes rendus de l'Académie des sciences* (Paris), volume 26, pages 535–538. 1848. http://gallica.bnf.fr/ark:/12148/bpt6k2983p/f539.item.r=.zoom

—Pasteur, L. 'Sur les relations qui peuvent exister entre la forme cristalline, la composition chimique et le sens de la polarisation rotatoire.' ('On the relations that can exist between crystalline form, and chemical composition, and the sense of rotary polarization.') *Annales de Chimie et de Physique*, 3rd series, volume 24, number 6, pages 442–459. 1848. https://books.google.com/books?id=gJ45AAAAcAAJ&pg=PA442&lpg=PA442#v=onepage

—Pasteur, L. 'Researches on the molecular asymmetry of natural organic products.' English translation of French original (1848), published by Alembic Club Reprints (volume 14, pages 1–46) in 1905, facsimile reproduction by SPIE in a 1990 book.

4 Flack, H. D. 'Louis Pasteur's discovery of molecular chirality and spontaneous resolution in 1848, together with a complete review of his crystallographic and chemical work.' *Acta Crystallographica Section A: Foundations and Advances*, volume 65, issue 5, pages 371–389. 2009. https://doi.org/10.1107/S0108767309024088

5 https://www.nobelprize.org/prizes/chemistry/1901/summary/

6 Van 't Hoff, J. H. 'Voorstel tot Uitbreiding der Tegenwoordige in de Scheikunde gebruikte Structuurformules in de Ruimte, benevens een daarmee samenhangende Opmerking omtrent het Verband tusschen Optisch Actief Vermogen en chemische Constitutie van Organische Verbindingen.' ('Proposal for the Extension of Current Chemical Structural Formulas into Space, together with Related Observation on the Connection between Optically Active Power and the Chemical Constitution of Organic Compounds.') Pamphlet published by the author. 1874. Available in English at: http://www.chemteam.info/Chem-History/Van't-Hoff-1874.html

—Le Bel, J. A. 'Sur les relations qui existent entre les formules atomiques des corps organiques et le pouvoir rotatoire de leurs dissolutions.' ('On the relations that exist between the atomic formulas of organic substances and the rotatory power of their solutions.') *Bulletin de la Société Chimique de Paris*, volume 22, pages 337–347. 1874. Online in French at: https://babel.hathitrust.org/cgi/pt?id=hvd.hc1j13;view=1up;seq=345

7 Joyce, G. F.; Schwartz, A. W.; Miller, S. L.; Orgel, L. E. 'The case for an ancestral genetic system involving simple analogues of the nucleotides.' *Proceedings of the National Academy of Sciences*, volume 84, issue 13, pages 4398–4402. 1987. https://doi.org/10.1073/pnas.84.13.4398

8 Nabarro, F. R. N.; Nye, J. F. 'Sir (Frederick) Charles Frank, O. B. E. 6 March 1911 – 5 April 1998.' *Biographical Memoirs of Fellows of the Royal Society*, volume 46, pages 177–196. 2000. https://doi.org/10.1098/rsbm.1999.0079

9 Frank, F. C. 'On spontaneous asymmetric synthesis.' *Biochimica et Biophysica Acta*, volume 11, issue 4, pages 459–463. 1953. https://doi.org/10.1016/0006-3002(53)90082-1

10 Soai, K.; Shibata, T.; Morioka, H.; Choji, K. 'Asymmetric autocatalysis and amplification of enantiomeric excess of a chiral molecule.' *Nature*, volume 378, issue 6559, pages 767–768. 1995. https://doi.org/10.1038/378767a0

11 Blackmond, D. G.; McMillan, C. R.; Ramdeehul, S.; Schorm, A.; Brown, J. M. 'Origins of asymmetric amplification in autocatalytic alkylzinc additions.' *Journal of the American Chemical Society*, volume 123, issue 41, pages 10103–10104. 2001. http://doi.org/10.1021/ja0165133

12 Lee, T. D.; Yang, C. N. 'Question of Parity Conservation in Weak Interactions.' *Physical Review*, volume 104, issue 1, pages 254–258. 1956. https://doi.org/10.1103/PhysRev.104.254

13 Wu, C. S.; Ambler, E.; Hayward, R. W.; Hoppes, D. D.; Hudson, R. P. 'Experimental Test of Parity Conservation in Beta Decay.' *Physical Review*, volume 105, issue 4, pages 1413–1415. 1957. https://doi.org/10.1103/PhysRev.105.1413

14 Yamagata, Y. 'A hypothesis for the asymmetric appearance of biomolecules on earth.' *Journal of Theoretical Biology*, volume 11, issue 3, pages 495–498. 1966. https://doi.org/10.1016/0022-5193(66)90110-X

15 Kondepudi, D. K.; Nelson, G. W. 'Weak neutral currents and the origin of biomolecular chirality.' *Nature*, volume 314, issue 6010, pages 438–441. 1985. https://doi.org/10.1038/314438a0

16 Quack, M. 'How Important is Parity Violation for Molecular and Biomolecular Chirality?' *Angewandte Chemie*, volume 41, issue 24, pages 4618–4630. 2002. https://doi.org/10.1002/anie.200290005

17 Kondepudi, D. K.; Kaufman, R. J.; Singh, N. 'Chiral Symmetry Breaking in Sodium Chlorate Crystallization.' *Science*, volume 250, issue 4983, pages 975–976. 1990. https://doi.org/10.1126/science.250.4983.975

18 Blackmond, D. G. 'The Origin of Biological Homochirality.' *Cold Spring Harbor Perspectives in Biology*, volume 2, issue 5, article a002147. 2010.

https://dx.doi.org/10.1101/cshperspect.a002147

19 Viedma, C. 'Chiral Symmetry Breaking During Crystallization: Complete Chiral Purity Induced by Nonlinear Autocatalysis and Recycling.' *Physical Review Letters*, volume 94, issue 6, article 065504. 2005. https://doi.org/10.1103/PhysRevLett.94.065504

20 Noorduin, W. L.; Izumi, T.; Millemaggi, A.; Leeman, M.; Meekes, H.; Enckevort, W. J.; Kellogg, R. M.; Kaptein, B.; Vlieg, E.; Blackmond, D. G. 'Emergence of a Single Solid Chiral State from a Nearly Racemic Amino Acid Derivative.' *Journal of the American Chemical Society*, volume 130, issue 4, pages 1158–1159. 2008. https://doi.org/10.1021/ja7106349

21 Viedma, C.; Ortiz, J. E.; de Torres, T.; Izumi, T.; Blackmond, D. G. 'Evolution of Solid Phase Homochirality for a Proteinogenic Amino Acid.' *Journal of the American Chemical Society*, volume 130, issue 46, pages 15274–15275. 2008. https://doi.org/10.1021/ja8074506

22 Morowitz, H. J. 'A mechanism for the amplification of fluctuations in racemic mixtures.' *Journal of Theoretical Biology*, volume 25, issue 3, pages 491–494. 1969. https://doi.org/10.1016/S0022-5193(69)80035-4

23 Klussmann, M.; Iwamura, H.; Mathew, S. P.; Wells, D. H. Jr; Pandya, U.; Armstrong, A.; Blackmond, D. G. 'Thermodynamic control of asymmetric amplification in amino acid catalysis.' *Nature*, volume 441, issue 7093, pages 621–623. 2006. https://doi.org/10.1038/nature04780
—Klussmann, M.; White, A. J. P.; Armstrong, A.; Blackmond, D. G. 'Rationalization and Prediction of Solution Enantiomeric Excess in Ternary Phase Systems.' *Angewandte Chemie International Edition*, volume 45, issue 47, pages 7985–7989. 2006. https://doi.org/10.1002/anie.200602520

24 Ball, P. 'Giving life a hand.' *Chemistry World*, issue 4, pages 30–31. 2007. https://www.chemistryworld.com/news/giving-life-a-hand/3001809.article

25 Tassinari, F.; Steidel, J.; Paltiel, S.; Fontanesi, C.; Lahav, M.; Paltiel, Y.; Naaman, R. 'Enantioseparation by crystallisation using magnetic substrates.' *Chemical Science*, advance article. 2019. https://doi.org/10.1039/C9SC00663J

26 Cronin, J. R.; Pizzarello, S. 'Enantiomeric Excesses in Meteoritic Amino Acids.' *Science*, volume 275, issue 5302, pages 951–955. 1997. https://doi.org/10.1126/science.275.5302.951

27 Glavin, D. P.; Dworkin, J. P. 'Enrichment of the amino acid l-isovaline by aqueous alteration on CI and CM meteorite parent bodies.' *Proceedings of the National Academy of Sciences*, volume 106, issue 14, pages 5487–5492. 2009. https://doi.org/10.1073/pnas.0811618106

28 McGuire, B. A.; Carroll, P. B.; Loomis, R. A.; Finneran, I. A.; Jewell, P. R.; Remijan, A. J.; Blake, G. A. 'Discovery of the interstellar chiral molecule propylene oxide (CH_3CHCH_2O).' *Science*, volume 352, issue 6292, pages

1449–1452. 2016. https://doi.org/10.1126/science.aae0328

29 Blackmond, D. G. 'The origin of biological homochirality.' *Cold Spring Harbor Perspectives in Biology*, volume 11, issue 3, article a032540. 2019. http://doi.org/10.1101/cshperspect.a032540

30 Hein, J. E.; Tse, E.; Blackmond, D. G. 'A route to enantiopure RNA precursors from nearly racemic starting materials.' *Nature Chemistry*, volume 3, issue 9, pages 704–706. 2011. https://doi.org/10.1038/nchem.1108
—Hein, J. E.; Blackmond, D. G. 'On the Origin of Single Chirality of Amino Acids and Sugars in Biogenesis.' *Accounts of Chemical Research*, volume 45, issue 12, pages 2045–2054. 2012. https://doi.org/10.1021/ar200316n
—Wagner, A. J.; Zubarev, D. Y.; Aspuru-Guzik, A.; Blackmond, D. G. 'Chiral Sugars Drive Enantioenrichment in Prebiotic Amino Acid Synthesis.' *ACS Central Science*, volume 3, issue 4, pages 322–328. 2017. https://doi.org/10.1021/acscentsci.7b00085

31 Morigaki, K.; Dallavalle, S.; Walde, P.; Colonna, S.; Luisi, P. L. 'Autopoietic Self-Reproduction of Chiral Fatty Acid Vesicles.' *Journal of the American Chemical Society*, volume 119, issue 2, pages 292–301. 1997. http://doi.org/10.1021/ja961728b

32 Saghatelian, A.; Yokobayashi, Y.; Soltani, K.; Ghadiri, M. R. 'A chiroselective peptide replicator.' *Nature*, volume 409, issue 6822, pages 797–801. 2001. https://doi.org/10.1038/35057238

33 Joshi, P. C.; Aldersley, M. F.; Ferris, J. P. 'Homochiral Selectivity in RNA Synthesis: Montmorillonite-catalyzed Quaternary Reactions of D, L-Purine with D, L-Pyrimidine Nucleotides.' *Origins of Life and Evolution of Biospheres*, volume 41, issue 3, pages 213–236. 2011. https://doi.org/10.1007/s11084-010-9222-1

34 Sczepanski, J. T.; Joyce, G. F. 'A cross-chiral RNA polymerase ribozyme.' *Nature*, volume 515, issue 7527, pages 440–442. 2014. https://doi.org/10.1038/nature13900

35 Davies, P. *The 5th Miracle: The search for the origin and meaning of life*. 1999. Simon & Schuster.

Chapter 13: Return of the Blobs

1 Luisi, P. L.; Walde, P.; Oberholzer, T. 'Enzymatic RNA synthesis in self-reproducing vesicles: An approach to the construction of a minimal synthetic cell.' *Berichte der Bunsengesellschaft für physikalische Chemie*, volume 98, issue 9, pages 1160–1165. 1994. https://doi.org/10.1002/bbpc.19940980918
—Oberholzer, T.; Wick, R.; Luisi, P. L.; Biebricher, C. K. 'Enzymatic RNA Replication in Self-Reproducing Vesicles: An Approach to a Minimal Cell.'

Biochemical and Biophysical Research Communications, volume 207, issue 1, pages 250–257. 1995. https://doi.org/10.1006/bbrc.1995.1180

2 'Jack W. Szostak – Biographical.' NobelPrize.org. https://www.nobelprize.org/prizes/medicine/2009/szostak/biographical/

3 Marshall, M. 'The secret of how life on Earth began'. BBC Earth. 2016. http://www.bbc.com/earth/story/20161026-the-secret-of-how-life-on-earth-began

4 Szostak, J. W.; Bartel, D. P.; Luisi, P. L. 'Synthesizing life.' *Nature*, volume 409, issue 6818, pages 387–390. 2001. https://doi.org/10.1038/35053176

5 Eigen, M. 'Selforganization of matter and the evolution of biological macromolecules.' *Naturwissenschaften*, volume 58, issue 10, page 504. 1971. https://doi.org/10.1007/BF00623322

6 Hanczyc, M. M.; Fujikawa, S. M.; Szostak, J. W. 'Experimental Models of Primitive Cellular Compartments: Encapsulation, Growth, and Division.' *Science*, volume 302, issue 5645, pages 618–622. 2003. https://doi.org/10.1126/science.1089904

7 Chen, I. A.; Walde, P. 'From Self-Assembled Vesicles to Protocells.' *Cold Spring Harbor Perspectives in Biology*, volume 2, issue 7, article a002170. 2010. http://doi.org/10.1101/cshperspect.a002170

8 Chen, I. A.; Roberts, R. W.; Szostak, J. W. 'The Emergence of Competition Between Model Protocells.' *Science*, volume 305, issue 5689, pages 1474–1476. 2004. https://doi.org/10.1126/science.1100757

9 Mansy, S. S.; Szostak, J. W. 'Thermostability of model protocell membranes.' *Proceedings of the National Academy of Sciences*, volume 105, issue 36, pages 13351–13355. 2008. https://doi.org/10.1073/pnas.0805086105

10 Hanczyc, M. M.; Szostak, J. W. 'Replicating vesicles as models of primitive cell growth and division.' *Current Opinion in Chemical Biology*, volume 8, issue 6, pages 660–664. 2004. https://doi.org/10.1016/j.cbpa.2004.10.002

11 Zhu, T. F.; Szostak, J. W. 'Coupled Growth and Division of Model Protocell Membranes.' *Journal of the American Chemical Society*, volume 131, issue 15, pages 5705–5713. 2009. https://doi.org/10.1021/ja900919c

12 Zhu, T. F.; Adamala, K.; Zhang, N.; Szostak, J. W. 'Photochemically driven redox chemistry induces protocell membrane pearling and division.' *Proceedings of the National Academy of Sciences*, volume 109, issue 25, pages 9828–9832. 2012. https://doi.org/10.1073/pnas.1203212109

13 Budin, I.; Debnath, A.; Szostak, J. W. 'Concentration-Driven Growth of Model Protocell Membranes.' *Journal of the American Chemical Society*, volume 134, issue 51, pages 20812–20819. 2012. https://doi.org/10.1021/ja310382d

14 Szostak, J. W. 'The eightfold path to non-enzymatic RNA replication.' *Journal of Systems Chemistry*, volume 3, issue 2. 2012. https://doi.org/10.1186/1759-2208-3-2

15 Adamala, K.; Szostak, J. W. 'Nonenzymatic Template-Directed RNA Synthesis Inside Model Protocells.' *Science*, volume 342, issue 6162, pages 1098–1100. 2013. https://doi.org/10.1126/science.1241888

16 Jin, L.; Engelhart, A. E.; Zhang, W.; Adamala, K.; Szostak, J. W. 'Catalysis of Template-Directed Nonenzymatic RNA Copying by Iron(II).' *Journal of the American Chemical Society*, volume 140, issue 44, pages 15016–15021. 2018. https://doi.org/10.1021/jacs.8b09617

17 Joyce, G. F.; Szostak, J. W. 'Protocells and RNA Self-Replication.' *Cold Spring Harbor Perspectives in Biology*, volume 10, issue 9, article a034801. 2018. https://doi.org/10.1101/cshperspect.a034801

18 Interview with John Sutherland.

19 Budin, I.; Prywes, N.; Zhang, N.; Szostak, J. W. 'Chain-Length Heterogeneity Allows for the Assembly of Fatty Acid Vesicles in Dilute Solutions.' *Biophysical Journal*, volume 107, issue 7, pages 1582–1590. 2014. https://doi.org/10.1016/j.bpj.2014.07.067

—Jin, L.; Kamat, N. P.; Jena, S.; Szostak, J. W. 'Fatty Acid/Phospholipid Blended Membranes: A Potential Intermediate State in Protocellular Evolution.' *Small*, volume 14, issue 15, article 1704077. 2018. https://doi.org/10.1002/smll.201704077

20 Blain, J. C.; Szostak, J. W. 'Progress Toward Synthetic Cells.' *Annual Review of Biochemistry*, volume 83, pages 615–640. 2014. https://doi.org/10.1146/annurev-biochem-080411-124036

—Adamala, K. P.; Engelhart, A. E.; Szostak, J. W. 'Collaboration between primitive cell membranes and soluble catalysts.' *Nature Communications*, volume 7, article 11041. 2016. https://doi.org/10.1038/ncomms11041

21 Adamala, K.; Szostak, J. W. 'Competition between model protocells driven by an encapsulated catalyst.' *Nature Chemistry*, volume 5, issue 6, pages 495–501. 2013. https://doi.org/10.1038/nchem.1650

22 Kamat, N. P.; Tobé, S.; Hill, I. T.; Szostak, J. W. 'Electrostatic Localization of RNA to Protocell Membranes by Cationic Hydrophobic Peptides.' *Angewandte Chemie*, volume 54, issue 40, pages 11735–11739. 2015. https://doi.org/10.1002/anie.201505742

23 Szostak, J. W. 'The Narrow Road to the Deep Past: In Search of the Chemistry of the Origin of Life.' *Angewandte Chemie*, volume 56, issue 37, pages 11037–11043. 2017. https://doi.org/10.1002/anie.201704048

24 Szathmáry, E. 'Founder of systems chemistry and foundational theoretical biologist: Tibor Gánti (1933–2009).' *Journal of Theoretical Biology*, volume 381, pages 2–5. 2015. http://dx.doi.org/10.1016/j.jtbi.2015.04.037

25 Gánti, T. *Az élet princípiuma* (*The Principles of Life*). 1971. Gondolat, Budapest.

26 Szathmáry, E. 'The origin of replicators and reproducers.' *Philosophical Transactions of the Royal Society B: Biological Sciences*, volume 361, issue

1474, pages 1761–1776. 2006. https://dx.doi.org/10.1098/rstb.2006.1912
27 Gánti, T. 'Theoretical deduction of the function and structure of the genetic material.' *Biológia*, volume 22, pages 17–35. 1974 (in Hungarian).
28 Gánti, T. *A Theory of Biochemical Supersystems*. 1979. Akadémiai Kiadó, Budapest.
29 Szathmáry, E.; Smith, J. M. 'The major evolutionary transitions.' *Nature*, volume 374, issue 6519, pages 227–232. 1995. https://doi.org/10.1038/374227a0
30 Oró, J.; Lazcano, A. 'A minimal living system and the origin of a protocell.' *Advances in Space Research*, volume 4, issue 12, pages 167–176. 1984. https://doi.org/10.1016/0273-1177(84)90559-3
31 Pascal, R.; Boiteau, L. 'Energy flows, metabolism and translation.' *Philosophical Transactions of the Royal Society B: Biological Sciences*, volume 366, issue 1580, pages 2949–2958. 2011. https://doi.org/10.1098/rstb.2011.0135
32 Pascal, R. 'Suitable energetic conditions for dynamic chemical complexity and the living state.' *Journal of Systems Chemistry*, volume 3, issue 3. 2012. https://doi.org/10.1186/1759-2208-3-3
33 Wagner, N.; Pross, A.; Tannenbaum, E. 'Selection advantage of metabolic over non-metabolic replicators: a kinetic analysis.' *Biosystems*, volume 99, issue 2, pages 126–129. 2010. https://doi.org/10.1016/j.biosystems.2009.10.005
34 Arsène, S.; Ameta, S.; Lehman, N.; Griffiths, A. D.; Nghe, P. 'Coupled catabolism and anabolism in autocatalytic RNA sets.' *Nucleic Acids Research*, volume 46, issue 18, pages 9660–9666. 2018. https://doi.org/10.1093/nar/gky598
35 Bonfio, C.; Valer, L.; Scintilla, S.; Shah, S.; Evans, D. J.; Jin, L.; Szostak, J. W.; Sasselov, D. D.; Sutherland, J. D.; Mansy, S. S. 'UV-light-driven prebiotic synthesis of iron–sulfur clusters.' *Nature Chemistry*, volume 9, issue 12, pages 1229–1234. 2017. https://doi.org/10.1038/nchem.2817
36 La Scola, B.; Audic, S.; Robert, C.; Jungang, L.; de Lamballerie, X.; Drancourt, M.; Birtles, R.; Claverie, J-M.; Raoult, D. 'A Giant Virus in Amoebae.' *Science*, volume 299, issue 5615, page 2033. 2003. https://doi.org/10.1126/science.1081867
37 Schulz, F.; Yutin, N.; Ivanova, N. N.; Ortega, D. R.; Lee, T. K.; Vierheilig, J.; Daims, H.; Horn, M.; Wagner, M.; Jensen, G. J.; Kyrpides, N. C.; Koonin, E. V.; Woyke, T. 'Giant viruses with an expanded complement of translation system components.' *Science*, volume 356, issue 6333, pages 82–85. 2017. https://doi.org/10.1126/science.aal4657
38 Nasir, A.; Kim, K. M.; Caetano-Anolles, G. 'Giant viruses coexisted with the cellular ancestors and represent a distinct supergroup along with superkingdoms Archaea, Bacteria and Eukarya.' *BMC Evolutionary Biology*,

volume 12, article 156. 2012. https://doi.org/10.1186/1471-2148-12-156

Chapter 14: Just Messy Enough

1 Powner, M. W.; Gerland, B.; Sutherland, J. D. 'Synthesis of activated pyrimidine ribonucleotides in prebiotically plausible conditions.' *Nature*, volume 459, issue 7244, pages 239–242. 2009. https://doi.org/10.1038/nature08013
2 Szostak, J. W. 'Systems chemistry on early Earth.' *Nature*, volume 459, issue 7244, pages 171–172. 2009. https://doi.org/10.1038/459171a
3 Kindermann, M.; Stahl, I.; Reimold, M.; Pankau, W. M.; von Kiedrowski, G. 'Systems chemistry: Kinetic and computational analysis of a nearly exponential organic replicator.' *Angewandte Chemie*, volume 44, issue 41, pages 6750–6755. 2005. https://doi.org/10.1002/anie.200501527
4 Deamer, D. W. 'Boundary structures are formed by organic components of the Murchison carbonaceous chondrite.' *Nature*, volume 317, issue 6040, pages 792–794. 1985. https://doi.org/10.1038/317792a0
5 Pizzarello, S.; Shock, E. 'The Organic Composition of Carbonaceous Meteorites: The Evolutionary Story Ahead of Biochemistry.' *Cold Spring Harbor Perspectives in Biology*, volume 2, issue 3, article a002105. 2010. https://doi.org/10.1101/cshperspect.a002105
6 Martins, Z.; Botta, O.; Fogel, M. L.; Sephton, M. A.; Glavin, D. P.; Watson, J. S.; Dworkin, J. P.; Schwartz, A. W.; Ehrenfreund, P. 'Extraterrestrial nucleobases in the Murchison meteorite.' *Earth and Planetary Science Letters*, volume 270, issues 1–2, pages 130–136. 2008. https://doi.org/10.1016/j.epsl.2008.03.026
7 Saladino, R.; Mincione, E.; Crestini, C.; Negri, R.; Di Mauro, E.; Costanzo, G. 'Mechanism of Degradation of Purine Nucleosides by Formamide. Implications for Chemical DNA Sequencing Procedures.' *Journal of the American Chemical Society*, volume 118, issue 24, pages 5615–5619. 1996. https://doi.org/10.1021/ja953527y
8 Saladino, R.; Crestini, C.; Pino, S.; Costanzo, G.; Di Mauro, E. 'Formamide and the origin of life.' *Physics of Life Reviews*, volume 9, issue 1, pages 84–104. 2012. https://doi.org/10.1016/j.plrev.2011.12.002
9 McGuire, B. A. '2018 Census of Interstellar, Circumstellar, Extragalactic, Protoplanetary Disk, and Exoplanetary Molecules.' *The Astrophysical Journal Supplement Series*, volume 239, number 2. 2018. https://doi.org/10.3847/1538-4365/aae5d2
10 Harada, K. 'Formation of Amino-acids by Thermal Decomposition of Formamide-Oligomerization of Hydrogen Cyanide.' *Nature*, volume 214, pages 479–480. 1967. https://doi.org/10.1038/214479a0
11 Saladino, R.; Crestini, C.; Costanzo, G.; Negri, R.; Di Mauro, E. 'A possible

prebiotic synthesis of purine, adenine, cytosine, and 4(3H)-pyrimidinone from formamide: implications for the origin of life.' *Bioorganic & Medicinal Chemistry*, volume 9, issue 5, pages 1249–1253. 2001. https://doi.org/10.1016/S0968-0896(00)00340-0

12 Saladino, R.; Ciambecchini, U.; Crestini, C.; Costanzo, G.; Negri, R.; Di Mauro, E. 'One-Pot TiO2-Catalyzed Synthesis of Nucleic Bases and Acyclonucleosides from Formamide: Implications for the Origin of Life.' *ChemBioChem*, volume 4, issue 6, pages 514–521. 2003. https://doi.org/10.1002/cbic.200300567

13 Saladino, R.; Crestini, C.; Ciambecchini, U.; Ciciriello, F.; Costanzo, G.; Di Mauro, D. 'Synthesis and Degradation of Nucleobases and Nucleic Acids by Formamide in the Presence of Montmorillonites.' *ChemBioChem*, volume 5, issue 11, pages 1558–1566. 2004. https://doi.org/10.1002/cbic.200400119

14 Saladino, R.; Crestini, C.; Cossetti, C.; Di Mauro, E.; Deamer, D. 'Catalytic effects of Murchison Material: Prebiotic Synthesis and Degradation of RNA Precursors.' *Origins of Life and Evolution of Biospheres*, volume 41, issue 5, article 437. 2011. https://doi.org/10.1007/s11084-011-9239-0

15 Saladino, R.; Carota, E.; Botta, G.; Kapralov, M.; Timoshenko, G. N.; Rozanov, A. Y.; Krasavin, E.; Di Mauro, E. 'Meteorite-catalyzed syntheses of nucleosides and of other prebiotic compounds from formamide under proton irradiation.' *Proceedings of the National Academy of Sciences*, volume 112, issue 21, pages E2746–E2755. 2015. https://doi.org/10.1073/pnas.1422225112

16 Saladino, R.; Di Mauro, E.; García-Ruiz, J. M. 'A Universal Geochemical Scenario for Formamide Condensation and Prebiotic Chemistry.' *Chemistry: A European Journal*, volume 25, issue 13, pages 3181–3189. 2018. https://doi.org/10.1002/chem.201803889

17 Ritson, D.; Sutherland, J. D. 'Prebiotic synthesis of simple sugars by photoredox systems chemistry.' *Nature Chemistry*, volume 4, issue 11, pages 895–899. 2012. https://doi.org/10.1038/nchem.1467

18 Patel, B. H.; Percivalle, C.; Ritson, D. J.; Duffy, C. D.; Sutherland, J. D. 'Common origins of RNA, protein and lipid precursors in a cyanosulfidic protometabolism.' *Nature Chemistry*, volume 7, issue 4, pages 301–307. 2015. https://doi.org/10.1038/nchem.2202

19 Xu, J.; Ritson, D. J.; Ranjan, S.; Todd, Z. R.; Sasselov, D. D.; Sutherland, J. D. 'Photochemical reductive homologation of hydrogen cyanide using sulfite and ferrocyanide.' *Chemical Communications*, volume 54, issue 44, pages 5566–5569. 2018. https://doi.org/10.1039/c8cc01499j

20 Woese, C. R. *The Genetic Code: The molecular basis for genetic expression*, page 189. 1967. Harper & Row.

21 Suárez-Marina, I.; Abul-Haija, Y. M.; Turk-MacLeod, R.; Gromski, P. S.;

Cooper, G. J. T.; Olivé, A. O.; Colón-Santos, S.; Cronin, L. 'Integrated synthesis of nucleotide and nucleosides influenced by amino acids.' *Communications Chemistry*, volume 2, article 28. 2019. https://doi.org/10.1038/s42004-019-0130-7

22 Petrov, A. S.; Bernier, C. R.; Hsiao, C.; Norris, A. M.; Kovacs, N. A.; Waterbury, C. C.; Stepanov, V. G.; Harvey, S. C.; Fox, G. E.; Wartell, R. M.; Hud, N. V.; Williams, L. D. 'Evolution of the ribosome at atomic resolution.' *Proceedings of the National Academy of Sciences*, volume 111, issue 28, pages 10251–10256. 2014. https://doi.org/10.1073/pnas.1407205111
—Petrov, A. S.; Gulen, B.; Norris, A. M.; Kovacs, N. A.; Bernier, C. R.; Lanier, K. A.; Fox, G. E.; Harvey, S. C.; Wartell, R. M.; Nud, N. V.; Williams, L. D. 'History of the ribosome and the origin of translation.' *Proceedings of the National Academy of Sciences*, volume 112, issue 50, pages 15396–15401. 2015. https://doi.org/10.1073/pnas.1509761112

23 Lanier, K. A.; Petrov, A. S.; Williams, L. D. 'The Central Symbiosis of Molecular Biology: Molecules in Mutualism.' *Journal of Molecular Evolution*, volume 85, issue 1–2, pages 8–13. 2017. https://doi.org/10.1007/s00239-017-9804-x

24 Monod, J. *Le hazard et la nécessité*. 1970. Éditions de Seuil, Paris. Published in English as *Chance and Necessity* by William Collins Sons & Co Ltd, 1972. Page 63–64.

25 Pross, A. *What Is Life? How chemistry becomes biology*. 2012. Oxford University Press. Pages 158–159.

26 Dworkin, J. P.; Deamer, D. W.; Sandford, S. A.; Allamandola, L. J. 'Self-assembling amphiphilic molecules: Synthesis in simulated interstellar/precometary ices.' *Proceedings of the National Academy of Sciences*, volume 98, issue 3, pages 815-819. 2001. https://doi.org/10.1073/pnas.98.3.815

27 Deamer, D. W. *Assembling Life: How can life begin on Earth and other habitable planets?* 2019. Oxford University Press.

28 Toppozini, L.; Dies, H.; Deamer, D. W.; Rheinstädter, M. C. 'Adenosine Monophosphate Forms Ordered Arrays in Multilamellar Lipid Matrices: Insights into Assembly of Nucleic Acid for Primitive Life.' *PLoS ONE*, volume 8, issue 5, e62810. 2013. https://doi.org/10.1371/journal.pone.0062810

29 Rajamani, S.; Vlassov, A.; Benner, S.; Coombs, A.; Olasagasti, F.; Deamer, D. W. 'Lipid-assisted Synthesis of RNA-like Polymers from Mononucleotides.' *Origins of Life and Evolution of Biospheres*, volume 38, issue 1, pages 57–74. 2008. https://doi.org/10.1007/s11084-007-9113-2

30 Olasagasti, F.; Kim, H. J.; Pourmand, N.; Deamer, D. W. 'Non-enzymatic transfer of sequence information under plausible prebiotic conditions.' *Biochimie*, volume 93, issue 3, pages 556–561. 2011. https://doi.org/10.1016/j.biochi.2010.11.012

31 Himbert, S.; Chapman, M.; Deamer, D. W.; Rheinstädter, M. C. 'Organization of Nucleotides in Different Environments and the Formation of Pre-Polymers.' *Scientific Reports*, volume 6, article 31285. 2016. https://doi.org/10.1038/srep31285
32 Black, R. A.; Blosser, M. C.; Stottrup, B. L.; Tavakley, R.; Deamer, D. W.; Keller, S. L. 'Nucleobases bind to and stabilize aggregates of a prebiotic amphiphile, providing a viable mechanism for the emergence of protocells.' *Proceedings of the National Academy of Sciences*, volume 110, issue 33, pages 13272–13276. 2013. https://doi.org/10.1073/pnas.1300963110
33 Namani, T.; Deamer, D. W. 'Stability of Model Membranes in Extreme Environments.' *Origins of Life and Evolution of Biospheres*, volume 38, issue 4, pages 329–341. 2008. https://doi.org/10.1007/s11084-008-9131-8
34 Cafferty, B. J.; Wong, A. S. Y.; Semenov, S. N.; Belding, L.; Gmür, S.; Huck, W. T. S.; Whitesides, G. M. 'Robustness, Entrainment, and Hybridization in Dissipative Molecular Networks, and the Origin of Life.' *Journal of the American Chemical Society*, volume 141, issue 20, pages 8289–8295. 2019. https://doi.org/10.1021/jacs.9b02554
35 Engelhart, A. E.; Powner, M. W.; Szostak, J. W. 'Functional RNAs exhibit tolerance for non-heritable 2'–5' versus 3'–5' backbone heterogeneity.' *Nature Chemistry*, volume 5, issue 5, pages 390–394. 2013. https://doi.org/10.1038/nchem.1623
36 Trevino, S. G.; Zhang, N.; Elenko, M. P.; Lupták, A.; Szostak, J. W. 'Evolution of functional nucleic acids in the presence of nonheritable backbone heterogeneity.' *Proceedings of the National Academy of Sciences*, volume 108, issue 33, pages 13492–13497. 2011. https://doi.org/10.1073/pnas.1107113108
37 Powner, M. W.; Zheng, S.-L.; Szostak, J. W. 'Multicomponent Assembly of Proposed DNA Precursors in Water.' *Journal of the American Chemical Society*, volume 134, issue 33, pages 13889–13895. 2012. https://doi.org/10.1021/ja306176n
38 Breaker, R. B.; Joyce, G. F. 'A DNA enzyme that cleaves RNA.' *Chemistry & Biology*, volume 1, issue 4, pages 223–229. 1994. https://doi.org/10.1016/1074-5521(94)90014-0
39 Szostak, J. W. 'An optimal degree of physical and chemical heterogeneity for the origin of life?' *Philosophical Transactions of the Royal Society B: Biological Sciences*, volume 366, issue 1580, pages 2894–2901. 2011. https://doi.org/10.1098/rstb.2011.0140
40 Fox, S. W. (ed) 'The Origins of Prebiological Systems and of their Molecular Matrices: Proceedings of a conference conducted at Wakulla Springs, Florida, on 27–30 October 1963 under the Auspices of the Institute for Space Biosciences, the Florida State University and the National Aeronautics and Space Administration', page 216. 1965. Elsevier, Inc.

41 Green, A. 'PSU professor who posted child pornography on his blog gets 2.5 years in prison.' *The Oregonian*, 8 July 2019. https://www.oregonlive.com/news/2019/07/psu-professor-who-posted-child-pornography-on-his-blog-gets-25-years-in-prison.html
42 Mizuuchi, R.; Lehman, N. 'Limited Sequence Diversity Within a Population Supports Prebiotic RNA Reproduction.' *Life*, volume 9, issue 1, page 20. 2012. https://doi.org/10.3390/life9010020
43 Reader, J. S.; Joyce, G. F. 'A ribozyme composed of only two different nucleotides.' *Nature*, volume 420, issue 6917, pages 841–844. 2002. https://doi.org/10.1038/nature01185
44 Morowitz, H. J. *Energy Flow in Biology: Biological organization as a problem in thermal physics*, pages 55–56. 1968. Academic Press, Inc.
45 Voytek, S. B.; Joyce, G. F. 'Niche partitioning in the coevolution of 2 distinct RNA enzymes.' *Proceedings of the National Academy of Sciences*, volume 106, issue 19, pages 7780–7785. 2009. https://doi.org/10.1073/pnas.0903397106
46 De León, L. F.; Podos, J.; Gardezi, T.; Herrel, A.; Hendry, A. P. 'Darwin's finches and their diet niches: the sympatric coexistence of imperfect generalists.' *Journal of Evolutionary Biology*, volume 27, issue 6, pages 1093–1104. 2014. https://doi.org/10.1111/jeb.12383
47 Woese, C. 'The universal ancestor.' *Proceedings of the National Academy of Sciences*, volume 95, issue 12, pages 6854–6859. 1998. https://doi.org/10.1073/pnas.95.12.6854
48 Woese, C. R. 'Interpreting the universal phylogenetic tree.' *Proceedings of the National Academy of Sciences*, volume 97, issue 15, pages 8392–8396. 2000. https://doi.org/10.1073/pnas.97.15.8392
49 Stewart, E. K. 'Growing unculturable bacteria.' *Journal of Bacteriology*, volume 194, issue 16, pages 4151–4160. 2012. https://doi.org/10.1128/JB.00345-12
50 Barras, C. 'We contain microbes so deeply weird they alter the very tree of life.' *New Scientist*, issue 3225. 2019. https://www.newscientist.com/article/mg24232250-200-we-contain-microbes-so-deeply-weird-they-alter-the-very-tree-of-life/
51 Quoted in Pross (2012), page 96.
52 Marshall, M. 'In the beginning: The full story of life on Earth can finally be told.' *New Scientist*, issue 3212. 2019. https://www.newscientist.com/article/mg24132120-300-in-the-beginning-the-full-story-of-life-on-earth-can-finally-be-told/
53 Roberts, N. M. W.; van Kranendonk, M. J.; Parman, S.; Clift, P. D. 'Continent formation through time.' In: *Continent Formation Through Time*, Geological Society, London, Special Publications, volume 389, pages 1–16. 2014. https://doi.org/10.1144/SP389.13

54 Van Kranendonk, M. J.; Philippot, P.; Lepot, K.; Bodorkos, S.; Pirajno, F. 'Geological setting of Earth's oldest fossils in the ca. 3.5 Ga Dresser Formation, Pilbara Craton, Western Australia.' *Precambrian Research*, volume 167, issues 1–2, pages 93–124. 2008. https://doi.org/10.1016/j.precamres.2008.07.003
—Djokic, T.; van Kranendonk, M. J.; Campbell, K. A.; Walter, M. R.; Ward, C. R. 'Earliest signs of life on land preserved in ca. 3.5 Ga hot spring deposits.' *Nature Communications*, volume 8, article 15263. 2017. https://doi.org/10.1038/ncomms15263

55 Deamer, D. W. 'The role of lipid membranes in life's origin.' *Life*, volume 7, issue 1. 2017. https://doi.org/10.3390/life7010005

56 Kompanichenko, V. N. 'Exploring the Kamchatka Geothermal Region in the Context of Life's Beginning.' *Life*, volume 9, issue 2, article 41. 2019. https://doi.org/10.3390/life9020041

57 Deamer, D. W.; Singaram, S.; Rajamani, S.; Kompanichenko, V.; Guggenheim, S. 'Self-assembly processes in the prebiotic environment.' *Philosophical Transactions of the Royal Society B: Biological Sciences*, volume 361, issue 1474, pages 1809–1818. 2006. https://doi.org/10.1098/rstb.2006.1905

58 Shapiro, R. 'Astrobiology: Life's beginnings.' *Nature*, volume 476, issue 7358, pages 30–31. 2011. https://doi.org/10.1038/476030a

59 Milshteyn, D.; Damer, B.; Havig, J.; Deamer, D. W. 'Amphiphilic Compounds Assemble into Membranous Vesicles in Hydrothermal Hot Spring Water but Not in Seawater.' *Life*, volume 8, issue 2, page 11. 2018. https://doi.org/10.3390/life8020011

60 'Scientists dunked test tubes in hot springs to recreate life's origins.' *New Scientist*, issue 3230. 2019. https://www.newscientist.com/article/2201961-scientists-dunked-test-tubes-in-hot-springs-to-recreate-lifes-origins/

61 Mulkidjanian, A. Y.; Bychkov, A. Y.; Dibrova, D. V.; Galperin, M. Y.; Koonin, E. V. 'Origin of first cells at terrestrial, anoxic geothermal fields.' *Proceedings of the National Academy of Sciences*, volume 109, issue 14, pages E821-E830. 2012. https://doi.org/10.1073/pnas.1117774109

62 Monnard, P-A.; Apel, C. L.; Kanavarioti, A.; Deamer, D. W. 'Influence of Ionic Inorganic Solutes on Self-Assembly and Polymerization Processes Related to Early Forms of Life: Implications for a Prebiotic Aqueous Medium.' *Astrobiology*, volume 2, issue 2, pages 139–152. 2004. https://doi.org/10.1089/15311070260192237

63 Sutherland J. D. 'The Origin of Life – Out of the Blue.' *Angewandte Chemie*, volume 55, issue 1, pages 104–121. 2015. https://doi.org/10.1002/anie.201506585

64 Ritson, D. J.; Battilocchio, C.; Ley, S. V.; Sutherland, J. D. 'Mimicking the surface and prebiotic chemistry of early Earth using flow chemistry.' *Nature*

Communications, volume 9, article 1821. 2018. https://doi.org/10.1038/s41467-018-04147-2

Epilogue: The Meaning of Life

1 Quoted on IMDb: https://www.imdb.com/title/tt0111281/characters/nm0209496
2 Klein, H. P.; Horowitz, N. H.; Levin, G. V.; Oyama, V. I.; Lederberg, J.; Rich, A.; Hubbard, J. S.; Hobby, G. L.; Straat, P. A.; Berdahl, B. J.; Carle, G. C.; Brown, F. S.; Johnson, R. D. 'The Viking Biological Investigation: Preliminary Results.' *Science*, volume 194, issue 4260, pages 99–105. 1976. https://doi.org/10.1126/science.194.4260.99
3 Klein, H. P. 'The Viking biology experiments: Epilogue and prologue.' *Origins of Life and Evolution of the Biosphere*, volume 21, issue 4, pages 255–261. 1992. https://doi.org/10.1007/BF01809861
4 McKay, D. S.; Gibson, E. K. Jr; Thomas-Keprta, K. L.; Vali, H.; Romanek, C. S.; Clemett, S. J.; Chillier, X. D. F.; Maechling, C. R.; Zare, R. N. 'Search for Past Life on Mars: Possible Relic Biogenic Activity in Martian Meteorite ALH84001.' *Science*, volume 273, issue 5277, pages 924–930. 1996. https://doi.org/10.1126/science.273.5277.924
5 Martel, J.; Young, D.; Peng, H.-H.; Wu, C.-Y.; Young, J. D. 'Biomimetic Properties of Minerals and the Search for Life in the Martian Meteorite ALH84001.' *Annual Review of Earth and Planetary Sciences*, volume 40, pages 167–193. 2012. https://doi.org/10.1146/annurev-earth-042711-105401
6 Kite, E. S.; Mayer, D. P.; Wilson, S. A.; Davis, J. M.; Lucas, A. S.; de Quay, G. S. 'Persistence of intense, climate-driven runoff late in Mars history.' *Science Advances*, volume 5, number 3, article eaav7710. 2019. https://doi.org/10.1126/sciadv.aav7710
7 Orosei, R.; Lauro, S. E.; Pettinelli, E.; Cicchetti, A.; Coradini, M.; Cosciotti, B.; Di Paolo, F.; Flamini, E.; Mattei, E.; Pajola, M.; Soldovieri, F.; Cartacci, M.; Cassenti, F.; Frigeri, A.; Giuppi, S.; Martufi, R.; Masdea, A.; Mitri, G.; Nenna, C.; Noschese, R.; Restano, M.; Seu, R. 'Radar evidence of subglacial liquid water on Mars.' *Science*, volume 361, issue 6401, pages 490–493. 2018. https://doi.org/10.1126/science.aar7268
8 Park, R. S.; Bills, B.; Buffington, B. B.; Folkner, W. M.; Konopliv, A. S.; Martin-Mur, T. J.; Mastrodemos, N.; McElrath, T. P.; Riedel, J. E.; Watkins, M. M. 'Improved detection of tides at Europa with radiometric and optical tracking during flybys.' *Planetary and Space Science*, volume 112, pages 10–14. 2015. https://doi.org/10.1016/j.pss.2015.04.005
9 Roth, L.; Saur, J.; Retherford, K. D.; Strobel, D. F.; Feldman, P. D.; McGrath, M. A.; Nimmo, F. 'Transient Water Vapor at Europa's South Pole.' *Science*,

volume 343, issue 6167, pages 171–174. 2014. https://doi.org/10.1126/science.1247051

10 Hand, K. P.; Carlson, R. W.; Chyba, C. F. 'Energy, Chemical Disequilibrium, and Geological Constraints on Europa.' *Astrobiology*, volume 7, number 6, pages 1006–1022. 2007. https://doi.org/10.1089/ast.2007.0156

11 https://solarsystem.nasa.gov/resources/10071/mountains-of-titan-map-2016-update/

12 Grasset, O.; Sotin, C.; Deschamps, F. 'On the internal structure and dynamics of Titan.' *Planetary and Space Science*, volume 48, issues 7–8, pages 617–636. 2000. https://doi.org/10.1016/S0032-0633(00)00039-8

13 Fortes, A. D. 'Exobiological Implications of a Possible Ammonia–Water Ocean inside Titan.' *Icarus*, volume 146, issue 2, pages 444–452. 2000. https://doi.org/10.1006/icar.2000.6400

14 Zhao, L.; Kaiser, R. I.; Xu, B.; Ablikim, U.; Ahmed, M.; Evseev, M. M.; Bashkirov, E. K.; Azyazov, V. N.; Mebel, A. M. 'Low-temperature formation of polycyclic aromatic hydrocarbons in Titan's atmosphere.' *Nature Astronomy*, volume 2, issue 12, pages 973–979. 2018. https://doi.org/10.1038/s41550-018-0585-y

15 Palmer, M. Y.; Cordiner, M. A.; Nixon, C. A.; Charnley, S. B.; Teanby, N. A.; Kisiel, Z.; Irwin, P. G. J.; Mumma, M. J. 'ALMA detection and astrobiological potential of vinyl cyanide on Titan.' *Science Advances*, volume 3, number 7, article e1700022. 2017. https://doi.org/10.1126/sciadv.1700022

16 McKay, C. P.; Smith, H. D. 'Possibilities for methanogenic life in liquid methane on the surface of Titan.' *Icarus*, volume 178, issue 1, pages 274–276. 2005. https://doi.org/10.1016/j.icarus.2005.05.018

17 Kawai, J.; Jagota, S.; Kaneko, T.; Obayashi, Y.; Yoshimura, Y.; Khare, B. N.; Deamer, D. W.; McKay, C. P.; Kobayashi, K. 'Self-assembly of tholins in environments simulating Titan liquidospheres: implications for formation of primitive coacervates on Titan.' *International Journal of Astrobiology*, volume 12, issue 4, pages 282–291. 2013. https://doi.org/10.1017/S1473550413000116

18 Deamer, D. W.; Damer, B. 'Can Life Begin on Enceladus? A Perspective from Hydrothermal Chemistry.' *Astrobiology*, volume 17, issue 9, pages 834–839. 2017. https://doi.org/10.1089/ast.2016.1610

19 Price, D. C.; Enrique, J. E.; Brzycki, B.; Croft, S.; Czech, D.; DeBoer, D.; DeMarines, J.; Foster, G.; Gajjar, V.; Gizani, N.; Hellbourg, G.; Isaacson, H.; Lacki, B.; Lebofsky, M.; MacMahon, D. H. E.; de Pater, I.; Siemion, A. P. V.; Werthimer, D.; Green, J. A.; Kaczmarek, J. F.; Maddalena, R. J.; Mader, S.; Drew, J.; Worden, S. P. 'The Breakthrough Listen Search for Intelligent Life: Observations of 1327 Nearby Stars over 1.10-3.45 GHz.' *arXiv*, 18 June 2019. https://arxiv.org/abs/1906.07750

20 Boyajian, T. S.; LaCourse, D. M.; Rappaport, S. A.; Fabrycky, D.; Fischer, D. A.; Gandolfi, D.; Kennedy, G. M.; Korhonen, H.; Liu, M. C.; Moor, A.; Olah, K.; Vida, K.; Wyatt, M. C.; Best, W. M. J.; Brewer, J.; Ciesla, F.; Csák, B.; Deeg, H. J.; Dupuy, T. J.; Handler, G.; Heng, K.; Howell, S. B.; Ishikawa, S. T.; Kovács, J.; Kozakis, T.; Kriskovics, L.; Lehtinen, J.; Lintott, C.; Lynn, S.; Nespral, D.; Nikbakhsh, S.; Schawinski, K.; Schmitt, J. R.; Smith, A. M.; Szabo, Gy.; Szabo, R.; Viuho, J.; Wang, J.; Weiksnar, A.; Bosch, M.; Connors, J. L.; Goodman, S.; Green, G.; Hoekstra, A. J.; Jebson, T.; Jek, K. J.; Omohundro, M. R.; Schwengeler, H. M.; Szewczyk, A. 'Planet Hunters IX. KIC 8462852 – where's the flux?' *Monthly Notices of the Royal Astronomical Society*, volume 457, issue 4, pages 3988–4004. 2016. https://doi.org/10.1093/mnras/stw218

21 Boyajian, T. S.; Alonso, R.; Ammerman, A.; Armstrong, D.; Asensio Ramos, A.; Barkaoui, K.; Beatty, T. G.; Benkhaldoun, Z.; Benni, P.; Bentley, R. O. et al. 'The First Post-Kepler Brightness Dips of KIC 8462852.' *The Astrophysical Journal Letters*, volume 853, number 1, article L8. 2018. https://doi.org/10.3847/2041-8213/aaa405

22 Extrasolar Planets Encyclopedia: http://exoplanet.eu/catalog/

23 Scharf, C. *The Copernicus Quest: The quest for our cosmic (in)significance*. 2014. Allen Lane.

24 Rimmer, P. B.; Xu, J.; Thompson, S. J.; Gillen, E.; Sutherland, J. D.; Queloz, D. 'The origin of RNA precursors on exoplanets.' *Science Advances*, volume 4, number 8, article eaar3302. 2018. https://doi.org/10.1126/sciadv.aar3302

25 Crane, L. 'We've found 4000 exoplanets but almost zero are right for life.' *New Scientist*, 22 March 2019. https://www.newscientist.com/article/2197406-weve-found-4000-exoplanets-but-almost-zero-are-right-for-life/

26 Trifonov, E. N. 'Vocabulary of Definitions of Life Suggests a Definition.' *Journal of Biomolecular Structure and Dynamics*, volume 29, issue 2, pages 259–266. 2012. https://doi.org/10.1080/073911011010524992

27 Morowitz, H. J.; Smith, E. 'Energy flow and the organization of life.' *Complexity*, volume 13, issue 1, pages 51–59. 2007. https://doi.org/10.1002/cplx.20191

28 Pross, A. *What Is Life? How chemistry becomes biology*. 2012. Oxford University Press.

29 Mullen, L. 'Defining Life: Q&A with Scientist Gerald Joyce.' *Space.com*, 1 August 2013. https://www.space.com/22210-life-definition-gerald-joyce-interview.html

30 Joyce, G. F. 'Foreword.' In: Deamer, D. W.; Fleischaker, G. R. *Origins of Life: The central concepts*. 1994. Jones and Bartlett Publishers, Boston (Massachusetts).

31 Benner, S. A. 'Defining life.' *Astrobiology*, volume 10, issue 10, pages 1021–1030. 2010. https://dx.doi.org/10.1089/ast.2010.0524
32 Cleland, C. E.; Chyba, C. F. 'Defining "Life".' *Origins of Life and Evolution of the Biosphere*, volume 32, issue 4, pages 387–393. 2002. https://doi.org/10.1023/A:1020503324273
33 Bruylants, G.; Bartik, K.; Reisse, J. 'Is it useful to have a clear-cut definition of life? On the use of fuzzy logic in prebiotic chemistry.' *Origins of Life and Evolution of Biospheres*, volume 40, issue 2, pages 137–143. 2010. https://doi.org/10.1007/s11084-010-9192-3
34 Bruylants, G.; Bartik, K.; Reisse, J. 'Prebiotic chemistry: A fuzzy field.' *Comptes Rendus Chimie*, volume 14, issue 4, pages 388–391. 2011. https://doi.org/10.1016/j.crci.2010.04.002
35 Pascal, R.; Pross, A.; Sutherland, J. D. 'Towards an evolutionary theory of the origin of life based on kinetics and thermodynamics.' *Open Biology*, volume 3, issue 11, article 130156. 2013. https://doi.org/10.1098/rsob.130156
36 Walker, S. I.; Packard, N.; Cody, G. D. 'Re-conceptualizing the origins of life.' *Philosophical Transactions of the Royal Society A: Mathematical, Physical and Engineering Sciences*, volume 375, issue 2109, article 20160337. 2017. https://doi.org/10.1098/rsta.2016.0337
37 Sutherland, J. D. 'Opinion: Studies on the origin of life – the end of the beginning.' *Nature Reviews Chemistry*, volume 1, article 0012. 2017. https://doi.org/10.1038/s41570-016-0012
38 Nisbet, H. B. *German Aesthetic and Literary Criticism: Winckelmann, Lessing, Hamann, Herder, Schiller, Goethe*. 1985. Cambridge University Press. https://books.google.co.uk/books?hl=en&lr=&id=xYg4AAAAIAAJ
39 Pross (2012), pages 185–186.

INDEX